LABORATORY TECHNIQUES IN ORGANIC CHEMISTRY

LABORATORY TECHNIQUES IN ORGANIC CHEMISTRY

V. K. AHLUWALIA
Visiting Professor
Dr. B.R. Ambedkar Centre for Biomedical Research
University of Delhi
Delhi

POOJA BHAGAT
Reader
Acharya Narendra Dev College
University of Delhi
Delhi

RENU AGGARWAL
Reader
Gargi College
University of Delhi
Delhi

TECHSAR™

NEW DELHI

Published and Printed by
Techsar Pvt. Ltd.
4435-36/7, Ansari Road, Daryaganj
New Delhi 110 002(India)
E-mail: tech_info@techsarworld.com
Website: www.techsarworld.com

ISBN 978-81-88237-32-6

Reprint 2025

Preface

This book deals with general information about work in Organic Chemistry Laboratory, viz., safety, first aid, different types of apparatus and their assemblies used for various types of reaction, stirring arrangements, heating techniques and low temperature experiments.

Various methods used for purification of organic compounds have been described. Besides the normal technique, the book includes write-up about molecular distillation, chromatography and electrophoresis.

Special emphasis has been given to the methods which can be used for working up of organic reactions. Various methods which can be used successfully for isolation of products from natural sources have been incorporated. Emphasis has also been given on the isolation of products from oily mixture using the technique of Liquid-Liquid extraction.

Methods for determining the criteria of purity of organic compounds have been discussed. The book also deals with drying and purification of solvents, preparation of spectroscopical grade solvents and HPLC solvents. The preparation of commonly used deuterated solvents (which are used for NMR spectroscopy work) is a special feature of this book.

V.K. Ahluwalia
Pooja Bhagat
Renu Aggarwal

Contents

1

General Information

This chapter provides general information for carrying out experiments in the organic laboratory. It describes the possible accidents in the laboratory and precautions to avoid them, the apparatus used in the laboratory for various reactions, viz. heating, cooling and stirring-techniques used in the laboratory, generation of gases and removal of solvents etc.

1.1 SAFETY IN THE LABORATORY

Chemistry laboratories are associated with several potential hazards and therefore certain precautions must be taken to minimize the probability of an accident. The knowledge of these hazards is the most important factor to avoid accidents and hence ensure safety in the laboratory.

1.1.1 General Safety Considerations

The following are the general factors, which must be adopted by an individual to avoid accidents:

(1) A thoughtless, careless and hurried behaviour must be avoided otherwise it may lead to an accident. Rather a thoughtful, careful, methodical approach must be adopted by an individual.

(2) The laboratory must always be kept neat and clean. Anything drops at bench or floor should be cleaned up immediately so as to avoid slipping.

(3) Running, eating, drinking or smoking in a laboratory should be avoided.

(4) Safety glasses (goggles) and laboratory coat must be worn at all time in the laboratory.

(5) Never work alone in the laboratory. The overnight experiments must be carried out in a special overnight room with necessary instructions.

(6) The position of first-aid box, exits and fire-extinguishers must be known by the person working in the laboratory.

(7) Chemicals should always be kept in their appropriate place in the shelves.

(8) Label should be read carefully before use. If the bottle is not labelled then either the content should be identified and then relabelled or it should be disposed of safely.

(9) Cracked or broken glass apparatus should not be used.

(10) Glass apparatus should also be stored at a proper place and not allowed to heap on the benches.

(11) The waste material should be disposed of regularly.

(12) Since most of the laboratory chemicals are poisonous, they should not be tasted and suction bulb should be used to suck them in filling pipettes.

(13) To observe the odour of the component, it should be kept at a safe distance from the face and gentle fan of the vapour should be done with the hand towards the nose.

(14) During heating, the mouth of the container should be away from the person and others nearby.

(15) Hot objects should not be picked up by hands. Tongs must be used.

(16) Before leaving the laboratory, the gas line should be closed properly.

(17) All the accidents must be reported to the Instructor immediately.

1.1.2 Fire

The most obvious hazard in the organic chemistry laboratory is the fire, as most of the liquids used, particularly solvents, are quite volatile and inflammable. The most common reasons for a fire are :

(a) Boiling of an inflammable solvent with a flame (without using condenser).

(b) Using the inflammable solvent near a burner.

(c) Using cracked apparatus, for inflammable liquids or cracking of the apparatus (containing inflammable liquids) while heating due to the faulty apparatus or unsuitable method of heating, and

(d) Adding pumice stones to a heated liquid, which is bumping violently, so the liquid comes out and catches fire. Therefore, always add pumice stones after cooling the bumping liquid for a few minutes.

When fire is large, go away from the burning area and bring it to the notice of the instructor immediately. The spread of fire can be prevented by removing the containers of inflammable solvents from the area and turning off any burner, if possible.

1.1.2.1 Burning clothing

If clothes catch fire, do not allow to run the victim as rapid movements fan the flames. The person should be made to lie down on the floor and fire proof blanket is wrapped around the burning clothes until the fire is extinguished.

In laboratories, sufficient number of fireproof blankets should be available. Each blanket should be kept in a labelled box (without lock) so that it can be taken out from the box without any delay. Alternatively the person whose clothes are set ablaze, should stand below a shower (which should be provided in the laboratory).

1.1.2.2 Bench fires

All the gas burners and the electrical hot plates near the area should be turned off and the inflammable material, if any, should be removed immediately.

Following methods are used to extinguish bench fires:

(A) A small fire can be caused by burning of a liquid in a flask, beaker or in an oil-bath. It can be extinguished by stopping the flow of air in the container, which can be done by covering the opening of it with a clean wet cloth or duster. And the compound or the reaction mixture can be recovered.

(B) **Dry Sand:** Most of the fires in the laboratory can be suppressed by the abundant use of dry sand. Laboratories should be equipped with buckets of dry sand properly distributed throughout and especially reserved for this purpose. After use, the sand should be thrown away and not reused as it is now contaminated with the inflammable, non-volatile materials.

So, for extinguishing fires, the sand is of great importance but it has some drawbacks also:

(a) The compound or reaction mixture is usually lost and cannot be recovered.

(b) The glass apparatus in that area may be broken down under the weight of the sand.

(C) **Carbon tetrachloride**: For small fires carbon tetrachloride spray may be used. Several sprayers are available which spray carbon tetrachloride under pressure. The fire is extinguished quickly by 'blanketing' effect of the heavy carbon tetrachloride vapour. While using carbon tetrachloride following points must be remembered:

(i) If sodium or potassium is present, it should not be used as it may result in violent explosions.

(ii) Highly poisonous phosgene vapours are formed when carbon tetrachloride is used, and, therefore, the laboratory must be ventilated immediately after extinguishing the fire to disperse these vapours.

(D) **Carbon dioxide:** The disadvantages of carbon tetrachloride can be overcome by using carbon dioxide under pressure. The fire extinguisher charged with carbon dioxide spray of the solid carbon dioxide. It can be used very effectively for extinguishing most of the fires in the laboratory, as :

(i) No adverse chemical reactions,

(ii) It has no electrical hazards, and

(iii) Minimal apparatus damage

Carbon dioxide fire extinguisher can be used more effectively by directing its nozzle towards the base of the flames. The fire extinguisher may be discharged by:

(a) Pulling out the pin after breaking a wire or plastic strip.

(b) Pointing the nozzle at the base of the flame from a distance of a few feet.

(c) Releasing the carbon dioxide snow by turning the valve open or squeezing the hand grip. Fire extinguisher should not be used violently, as fire can be made worse easily by allowing the strong flow of solid carbon dioxide from nozzle and the bottles and apparatus containing inflammable material can be damaged.

(E) **Dry Powder Extinguisher**: For organic solvents or burning oil, water should not be used as it will spread fire. In such cases, a mixture of sand and sodium bicarbonate is used. Dry powder extinguisher is available commercially and is effective for fires caused by magnesium and metal hydrides.

General precautions to be taken into consideration to avoid fire hazards, are already discussed in the Section 1.1.1, but the more specific precautions for the fire hazards are:

(I) Inflammable Materials: The inflammability of a liquid can be measured by its 'flash point' which is defined as the temperature at which the liquid gives rise to ignitable vapours. If the flash point of the liquid is less than 15°C, then it is considered highly inflammable. Table 1.1 shows the flash points of some commonly used inflammable solvents.

Table 1.1 Flash points of some of the inflammable solvents

S. No.	*Name of the Solvent*	*Flash Point (°C)*
1.	Light Petroleum (b.p. 40-60°C)	–49
2.	Diethyl ether	–45
3.	Carbon disulphide	–30*
4.	Light petroleum (b.p. 60-80°C)	–23
5.	Acetone	–18
6.	Benzene	–11
7.	Ethanol	12

*Autoignition temperature [the temperature at which the vapour will ignite in air spontaneously] of carbon disulphide is very low, 100°C and so can ignite on contact with boiling water baths. And, therefore, it is the most dangerous solvent in the laboratory and hence, should not be used until any other suitable solvent is not available.

So, inflammable materials should be handled with care. The vapours of such materials should not be allowed to escape in the laboratory as these vapours may travel over to a distant burner and get ignited. Besides most of these solvents are poisonous, benzene is carcinogenic and its use should be avoided. Some of the precautions for their use are :

(a) These materials should be kept in a tightly closed container.

(b) Small quantities of such materials should be stored at a time in the laboratory.

(c) These should not be vaporised in the atmosphere, rather should be condensed and collected.

(d) All the flames on the bench must be extinguished while working with these materials.

(e) Highly volatile solvents should be stored in the refrigerator [specially designed for laboratory and not the domestic one] in a tightly closed glass containers.

(II) Ignition Sources: Direct flames should be used in the laboratory rarely to avoid the fire hazards. The safe methods for heating are steam bath, oil bath, electric heating mantle etc. While using the flame, first a careful survey of the nearby compounds and apparatus must be done and the burner should be turned off after the use immediately.

The vapours of inflammable material may be ignited by sparking from electrical contacts and to avoid this, the laboratory should be properly ventilated so as to prevent the accumulation of the vapours.

(III) Leaks in Oxygen Cylinders: The oxygen cylinders should be tested regularly for leaks by brushing a dilute solution of detergent on the valve joints. The oxygen cylinders, which leak, should be taken out from the laboratory.

If the oxygen cylinder leaks when there is no ventilation in the laboratory, the amount of oxygen in the air will increase and may result in a violent fire when an ignition source is turned on in the close vicinity of inflammable solvent.

(IV) Sodium Metal: The sodium metal reacts violently with water to undergo a highly exothermic reaction and results in a fire and explosion hazard. These fires [fires due to sodium metal] are very hot and localised. In such cases, sand or dry powder extinguisher should be used. As carbon tetrachloride or carbon dioxide gives the violent explosions with sodium metal, and therefore they should not be used in such fires.

Precaution is required when dealing with used sodium wire for solvent drying purposes. It should not be thrown in the sink as such. The used sodium wire should be first covered with 2-propanol and left aside for at least 2 hours with occasional stirring until all the sodium is destroyed. Then these contents should be poured into excess of water and the bottle is washed a number of times with alcohol. Now the bottle can be washed with water safely.

(V) Metal hydrides: Metal hydrides e.g. Lithium aluminium hydride, sodium hydride also react fiercely with water to give hydrogen and results in a fire and explosion hazard. So care must be taken while handling with metal hydrides. When a metal hydride is used in a reaction, then excess of the metal hydride must be destroyed by adding ethyl acetate or acetone carefully.

First aid for burns from flames, hot plates, hot glass apparatus etc. is the cold water. Any kind of dressing should never be applied on it. If the burn is more than the small blister, call the doctor.

1.1.3 Explosions

Explosion is an exothermic reaction, which gets out of control and the container in which this reaction is carried out, is shattered off.

Explosive substances are:

(a) Compounds undergoing internal redox reactions e.g. polynitro compounds, trinitrotoluene, picric acid. These substances, when damp with water, are safe.

(b) Compounds, which can decompose to give very stable molecules e.g. acetylenes, nitrogen iodides, diazo compounds, diazonium salts, azides, peroxides.

Explosive mixtures are formed when a powerful oxidant, e.g. perchloric acid, permanganates, chromium trioxide, concentrated hydrogen peroxide, is mixed with easily oxidised organic substances, e.g. alcohols, carbohydrates etc.

The most common explosion hazards in the laboratory are:

(a) An exothermic reaction that gets out of control e.g. reaction of sodium metal and metal hydrides with water (already discussed in the previous Section).

(b) Explosions upon heating, drying or distillation of the unstable compounds e.g. diazonium compounds, diazonium salts etc.

(c) Explosions due to presence of peroxides in ether solvents.

(d) Reaction of perchloric acid with cork, wood, cloth, rubber is violent.

(e) Explosions during vacuum distillation or sublimation due to the cracking of the glass apparatus.

(f) Sodamide and potassium metal form oxide films by surface oxidation, which results in explosion during handling. Potassium can form this film even when stored under oil and may undergo explosion when this film is removed with a knife.

(g) Use of chromic acid and nitric acid as cleaning agents is extremely dangerous and may result in violent explosions when added to remove the remaining residues from the glass apparatus.

(h) Tollen's reagent forms explosive silver azides, when kept aside before use. And use of such Tollen's reagent is highly dangerous.

(i) For opening of glass ampoules of volatile chemicals, cooling is necessary. If cooling is not done carefully, the glass may crack and the chemical may enter into cooling bath and result in a violent explosion.

The following precautions should be considered to minimise the explosion hazards:

(a) Any explosive substance should not be used until a safe suitable method is available.

(b) Reaction of an explosive substance should be carried out on a small scale in a fuming cupboard with all the precautions.

(c) For an exothermic reaction or reaction with an explosive substance, the reagent should be added dropwise with rapid and continuous stirring. Each drop should be added after ensuring the complete reaction of each previously added drop of the reagent. In such cases, overcooling must be avoided as it may inhibit the reaction and the reagent may get accumulated in the reaction mixture so as temperature is then raised, an explosion may result. Instead, heating is done to ensure the completion of the reaction after adding each drop.

(d) The explosion of peroxides in ether solvents results due to the increase in the peroxide content residue [initially formed by exposure of ether to air and light]. Such explosions can be avoided by

- Storage of ethers in a dark glass container.
- They should not be stored for long periods.
- Peroxides must be removed from these solvents before they are distilled off.
- Presence of peroxides in ethers should be tested immediately before use as they rapidly give peroxide on exposure to air and light.

(e) The apparatus of a violent reaction should be surrounded by a safety shield [covered from all sides and open at the top]. Fuming cupboards should not be used as safety shield for violent reactions as the glass of these cupboards cannot withstand an explosion and may break down into pieces. The safety shield will give protection from flying glass and hot contents of the flask during an explosion.

(f) Potassium having heavy oxide film should not be used. It should be destroyed by adding excess of propan-2-ol carefully. Similarly yellow lumps of sodamide should not be used but destroyed by mixing with solid ammonium chloride.

(g) Chromic acid and nitric acid should not be used as cleaning agents.

(h) Always freshly prepared Tollens reagent should be used. If some of it is left behind, it should be destroyed by adding aqueous sodium chloride into it.

(i) Cooling of ampoules of highly reactive chemicals should be carried out carefully. For opening of ampoules, first make a clean scratch in its neck with the help of a sharp file and then crack it off by touching the molten end of a red-hot thin glass tube on the scratch.

Also avoid the resealing of the ampoules. The smaller size ampoules should be used so that all of its content can be used.

1.1.4 Poisoning

Poisoning by mouth is not common in the laboratories as (a) nothing should be eaten or drunk in the laboratory, (b) nothing e.g. milk, water etc. should be stored in the laboratory refrigerator, (c) All pipetting should be done with suction bulb and not by mouth, (d) Hands should be washed properly after working in the laboratory and before eating.

1.1.4.1 Solids/liquids poisoning

Even if it happens then the most suitable first aid treatment for poisoning depends on the nature of poison. The following are examples of the first aid treatment for different types of solids and liquids:

(a) **In the mouth but not swallowed:** If the solid/liquid has gone in the mouth but not swallowed then it must be spitted out at once and wash your mouth through water repeatedly.

(b) **If swallowed:** If it is swallowed then call a doctor immediately and meanwhile give an antidote depending upon the nature of the poison, e.g.

(i) *Acids:* First aid for acid poisoning is the dilution by drinking much water and then lime-water or milk of magnesia. No attempt should be made to make the patient vomit. Milk may be given in the last.

(ii) *Caustic alkalis:* For caustic alkali poisoning, it is also diluted by drinking much water and then the vinegar, orange or lemon juice, or solutions of citric acid or lactic acid. In the last, milk may be given. Here, also the precaution of no attempt to vomit should be taken into consideration.

(iii) *Heavy metal salts:* For heavy metal salts, milk or white of an egg should be given.

(iv) *Compounds of Arsenic or Mercury:* In such cases the patient is allowed to vomit and therefore, emetic should be given immediately, e.g. one teaspoon of mustard or zinc sulphate or salt in a glassful of warm water.

(v) *Alkyl cyanide:* For such poisoning, amyl nitrile from the freshly opened capsules is inhaled immediately.

Some other examples of highly toxic solids are inorganic cyanides, osmium tetraoxide, selenium and its compounds, thallium salts, vanadium pentoxide etc. So special care should be taken while handling them.

1.1.4.2 Gas poisoning

The very common way of poisoning is gas poisoning i.e. inhalation into the lungs. It involves the gases, vapours, dust etc. And, therefore when toxic powders, volatile liquids and harmful or inflammable gases (which involves almost all gases except oxygen, nitrogen, helium, neon or argon) are used, they should be used in the efficient fuming cupboards. Particular care should be taken for smelling unknown compounds for identification purposes.

In gas poisoning, the patient should be removed from the laboratory to the open air, and the clothing at the neck should be loosened. Artificial respiration should be applied, if breathing has stopped or become very weak until a doctor arrives.

In case of chlorine or bromine poisoning, inhale ammonia vapours and for soothing the throat suck eucalyptus pastilles or drink warm dilute peppermint or cinnamon essence.

1.1.5 Cuts

Most of the cuts occur while applying force to insert a glass tube or thermometer into a stopper or an adapter. In this attempt, the tube or thermometer may break and cut the palm or the base of the finger. So, first lubricate the stopper or adapter with glycerine, excess of which can be washed off with water and protect the hands with a towel. To avoid such accidents, the best way is to use ground glass jointed glassware.

1.1.5.1 Minor cuts

For a minor cut, wash thoroughly with disinfectant e.g. rectified spirit, dettol (diluted with water) etc. and then with water, remove glass, if any, with sterile forceps and apply sterile dressing [cotton wool soaked in a disinfectant followed by bandage].

1.1.5.2 Serious cuts

For serious cuts, call the doctor immediately and in the meantime, wash it with a disinfectant. Attempt should be made to minimise bleeding by applying pressure immediately above the cut. [It should be noted that the continuous pressure should not be applied for more than five minutes.]

If glass is present in the cut, wash thoroughly before applying pressure and remove it with sterile forceps. But if excessive bleeding is there, the application of pressure to minimise bleeding should be preferred than the removal of the glass.

1.1.6 Burns

In case of severe burns, call the doctor at once and in the meantime provide them with the first-aid in the laboratory.

(A) *By dry heat:* If burns are caused by heating through flames, hot objects etc. then;

  (a) For the burns in which skin is unaffected (not broken or become red), apply standard burn ointment e.g. acriflavine (burnol), tannic acid (tannafax).

  (b) For larger burns in which skin is broken or reddened or blistered, apply acriflavine or crystal violet jelly immediately and call the doctor at once.

(B) *By boiling water:* Apply acriflavine emulsion immediately.

(C) *By acids:* Wash the area immediately with large amounts of water thoroughly, followed by saturated sodium bicarbonate solution and then with the water again. For larger burns, apply disinfectant, dry the skin and then apply acriflavine emulsion.

(D) *By alkalis:* Wash the affected area immediately with lots of water, followed by 1% acetic acid and then finally with water. For serious burns, here also, apply disinfectant, dry the skin and then apply acriflavine emulsion.

(E) *By bromine:* While dealing with the experiments involving bromine, small bottles of light petroleum [b.p. 80-100°] should be available immediately. If the bromine is split on the skin, immediately wash the area with excess of petroleum, apply glycerine on it and rub it gently into the skin. After a few minutes, remove the excess of glycerine and then finally apply acriflavine emulsion.

(F) *By sodium:* Remove the solid metal, if any, in the affected area with the help of sterile forceps. Wash with water thoroughly followed by 1 percent acetic acid. Apply a cotton wool soaked in acriflavine emulsion or olive oil.

(G) *By dimethyl sulphate:* Wash the area immediately with lots of concentrated ammonia solution. Rub gently with a cotton wool soaked in the ammonia solution.

(H) *By phosphorus:* Wash with cold water thoroughly, and then with silver nitrate solution. In case of serious burns, wash again with water and apply acriflavine emulsion.

(I) *By organic substances:* Wash the affected area with rectified spirit followed by soap and warm water.

1.1.7 Eye Accidents

In eye accidents, doctor should be consulted in all cases. While if the accident is serious, the doctor should be called at once and the meantime first aid should be given.

First aid for chemicals in the eyes is to wash them thoroughly with water for several minutes by gentle flow of water into the eyes, i.e. by either using a beaker or by putting the head under the tap. Contact lens must be removed to wash the eyes. If the chemical has entered in the eyes then:

(a) *For acids:* For dilute acids, wash the eyes with 1 percent sodium bicarbonate solution repeatedly with the help of eye cup. While for concentrated acids, first wash the eyes with a liberal quantity of water followed by 1 percent sodium bicarbonate solution.

(b) *For alkalis:* For dilute alkalis, wash the eyes with 1 percent boric acid solution repeatedly. While for concentrated alkalis, first wash the eyes with large volume of water and then by 1 percent boric acid solution.

(c) *For bromine:* First wash the eyes with water and then immediately with 1 percent sodium bicarbonate solution.

If the glass has entered the eye, remove the loose glass by washing the eye gently with water or with the forceps. The imbedded glass should not be removed as it may cause more damage to eyes. Call the doctor immediately.

In minor eye accidents, soreness of the eye may be relieved by adding 1 drop of castor oil in the cover of the eye.

Eyes are more prone to accidents by flying particles of glass etc. or splashing of chemicals. And, therefore, safety goggles should be worn at all times in the laboratory, especially while dealing with dangerous operations e.g. vacuum distillation of large volumes of inflammable solvents, using liquid gases etc.

1.1.8 Electrical Safety

In chemical laboratories, we generally talk about the accidents caused by the corrosive and toxic reagents. However, the accidents may also be caused by the mishandling and malfunctioning of the electrical appliances. The probability of these accidents can be reduced by taking few things into account while handling them.

While dealing with electrical equipment, the apparatus should be set up on a dry bench. And the best way of handling them is to first set up the apparatus and then switch it on. Whenever, adjustment is required, it should be done by switching off the apparatus first.

Before using these equipment, the following factors should be taken into consideration:

1. It is earthed properly.
2. No loose wires or connections, whenever, the power source is away from the apparatus set up. The wire should be secured with adhesive tape run along the side of a bench.
3. The cable of the electrical equipment should have good insulation.
4. It should be protected by a correct rating fuse.
5. If any chemical is spilled on these appliances, they should be used only after thorough cleaning and drying.

Even if by taking all the above precautions, the electrical shock happens, then first switch the plug off. In a state of shock, i.e. faint/collapsed/pale/sweating etc., lie the victim flat with the legs raised to one foot, loosen the clothes around neck. Keep warm by using blanket. And call the doctor immediately.

1.1.9 Ultraviolet Radiation

Sources which emit ultraviolet radiations should not be viewed directly as it may cause eye damage. Burns on skin are developed on exposure to intense U.V. radiation and may result in extensive tissue damage on prolonged exposure to skin. Following precautions should be taken into account while handling with U.V. radiation sources to avoid the above damages:

(a) Close-fitted goggles which are specially meant for such purposes (opaque to U.V. radiation) should be used.

(b) Apparatus should be placed in protective screens.

(c) Gloves should be used to avoid the risk of exposure.

(d) There should be a proper ventilation to avoid the excessive concentration of ozone, which is highly irritant and toxic, formed when oxygen is irradiated with U.V. radiation in 185 nm region.

1.1.10 List of Toxic and Hazardous Chemicals Along with their Properties

Most of the compounds used in organic chemistry laboratory are toxic and hazardous. Toxicity is the property of a compound, which causes injury on reaching a susceptible site. These compounds can enter into the body by inhalation (into the lungs), ingestion (through the mouth) or absorption by skin. And, therefore, they should be handled carefully, so that their entry into the body can be prevented. A brief description of the hazardous properties of some generally used chemicals are listed below. The following list will help in knowing the dangers by substances often found in an organic laboratory.

Acetaldehyde: It is inflammable. On inhalation, its vapours result in the irritation of eyes, skin and respiratory organs. It should be stored in a cool place.

Acetic acid: First aid for the spills caused by its touching on the skin is to dilute it with large amounts of water.

Acetic anhydride: Pungent smell corrosive liquid. It decomposes slowly with water to form acetic acid. Its vapours result in the irritation of eyes, skin and mucous membrane and causes nausea. It also causes blisters in the skin if not washed off immediately. First aid for spill on the skin is to wash it off with water followed by dilute ammonia solution.

Acetonitrile: It is flammable and on inhalation causes acute dizziness, headache and nausea.

Acetyl chloride: It is a fuming liquid, which reacts violently with water to form acetic acid and hydrogen chloride (HCl). It is highly irritant and corrosive and causes inflammation of skin. When spilled on skin, wash it with water followed by dilute ammonia solution. It must be stored in a cool, well-ventilated room.

Acid chlorides: e.g. Acetyl chloride, benzoyl chloride, benzene-sulphonyl chloride, p-toluene-sulphonyl chloride, all of them are highly corrosive and first aid, if they spill on the skin is first wash with water and finally with dilute ammonia solution.

Acids: Corrosive and if spill, then wash off with large amount of water and then finally with sodium bicarbonate solution.

Acrolein: It is a flammable and pungent liquid. On inhalation, its vapours cause inflammation of eye, nasal, throat and skin. It should be kept in a fuming cupboard.

Aluminium chloride: It is corrosive and reacts with water violently to give hydrogen chloride.

Ammonia: The gas is irritating in odour and on inhalation causes suffocation, nausea, cough, bronchitis and irritation to the skin. First aid for inhalation is breathing of fresh air, steam. While if vapours enter the eyes, wash them with water for 15 minutes.

Aniline: On inhalation causes dizziness, nausea and abdominal pain. Through intact skin, fatal doses can be absorbed.

Benzene: It is a highly inflammable liquid. It causes euphoria, headache and narcosis. Its chronic toxicity includes bone marrow depression and acute anaemia, which may lead to leukaemia.

Benzoyl peroxide: It is highly explosive especially, if heated and so it should not be heated for melting point determination or recrystallization.

Bromine: It is a irritating and highly corrosive liquid. On inhalation, it causes dizziness, headache, bronchitis, nausea and skin burns. If inhaled, then inhale first fresh air and then the vapours of a very dilute solution of ammonia. Great care is required while handling bromine :

wearing gloves and laboratory apron. When spilled on skin, first aid is to wash instantly with water, rinse with ethanol and rub in glycerine.

n-Butanol: It has a moderate fire risk. On inhalation, causes dizziness, paralysis and respiratory inflammation.

n-Butyl acetate: On inhalation it causes cough, conjunctivitis and headache and loss of appetite.

n-Butyl lithium: It is a highly toxic and irritant liquid, which on contact with moist air ignites and should be preserved below 15°C.

Carbon disulphide: It is a highly inflammable liquid, which causes abdominal pain, headache and vomiting.

Carbon tetrachloride: It causes poisoning by inhalation, ingestion or absorption through skin. It causes irritation of eyes, headache, abdominal cramps and nervousness.

Chlorine: It is a highly corrosive, suffocating gas. On inhalation it causes irritation of eyes, difficult breathing, cough, nausea and pain. Its first aid for inhalation and spills on skin is same as that of bromine.

Chloroacetyl chloride: It causes irritation of eyes, nasal and throat.

Chloroform: It is a volatile and potentially harmful liquid, which causes unconsciousness, shortness of breath and vomiting.

Chlorosulphonic acid: It is a highly corrosive liquid, which reacts explosively with water to form sulphuric acid and hydrogen chloride. If spills on skin or cloth, wash it with plenty of water.

Diazomethane: It is a highly toxic and irritant liquid, which decomposes violently with water and alcohol. It possesses a highly explosion hazard.

1,2-Dichloroethane: It is a flammable liquid, which causes anxiety, convulsions, headache, weakness and irritation of respiratory tract.

Dichloromethane: It is also called as methylene chloride. It is nacrotic in high concentrations and potentially harmful on prolonged exposure to small quantities.

Diethyl ether: It is a highly volatile and flammable liquid. It can reach to the source of ignition even by travelling a long distance. On inhalation it causes headache, paralysis and irritation of respiratory tract. It should be stored in a cool place.

Dimethyl sulphate: It is highly dangerous as can be absorbed in fatal doses through intact skin immediately. If spilles on skin, wash it off with dilute ammonia solution and remove contaminated cloth.

Dimethyl sulphoxide: It is not harmful in itself. But absorbed quickly through intact skin and then help in absorption of other materials through skin.

Ethers: e.g. ethyl ether, isopropyl ether, tetrahydrofuran, dioxane, they are highly inflammable when stored, in presence of oxygen, form explosive peroxides, and therefore, ethers should be stored in full, airtight bottles, otherwise should be discarded within two months. Whenever the concentration of large amounts of ether is required, first the absence of the peroxides should be checked. Always take a precaution that never concentrate the ether to dryness by using a flame.

Detection of peroxides in ethers or hydrocarbons: Add 0.5-1 ml of the compound to be tested to equal volume of glacial acetic acid and 100 mg of sodium or potassium iodide. Yellow colouration indicates a low concentration of peroxides and a brown colour indicates a high concentration of peroxides.

Removal of Peroxides from ethers: Shake the ether with small portions of a solution made by 60 g of ferrous sulphate dissolved in 6 ml of concentrated sulphuric acid in 100 ml of water.

Ethyl bromide: It is potentially harmful on prolonged exposure to small quantities.

Formaldehyde: A highly reactive, pungent smell gas which causes conjunctivitis, nausea and irritation of skin.

Fuming Nitric acid: It is 95% nitric acid containing oxides of nitrogen. It is highly corrosive and if spills, dilute with plenty of water followed by sodium bicarbonate solution.

Fuming Sulphuric acid: It is oleum i.e. concentrated sulphuric acid containing dissolved sulphur trioxide. It is also highly corrosive and when spills, wash it first with excess of water and then finally with sodium bicarbonate solution.

Halogens: All halogens are corrosive and toxic. As fluorine reacts violently with large number of compounds, proper care should be taken while handling.

Halogenated solvents: e.g. 1,1,2,2-Tetrachloroethane, chloroform, carbontetrachloride etc. 1,1,2,2-tetrachloroethane is the most dangerous halogenated solvents as it can be absorbed quickly through the skin.

Hydrazine: It is a fuming and highly explosive liquid with harmful vapours. It is dangerous when used with oxidising agents. It causes conjunctivitis, nausea and irritation of skin.

Hydrides: e.g. Lithium aluminium hydride, sodium hydride. They explosively react with water and liberate igniting hydrogen. And, therefore, their contact with water must be avoided. Calcium hydride reacts less vigorously with water. Borohydrides also react less vigorously with water but quickly with acidic solutions to liberate hydrogen.

Hydrobromic acid: It is highly corrosive and irritant. If spills, wash with excess of water followed by dilute sodium bicarbonate solution.

Hydrochloric acid: Same as hydrobromic acid.

Hydrofluoric acid: It reacts instantly with glass and rapidly destroys organic tissues. New thick plastic gloves (having no holes) should be worn, while handling it. If burns on skin, immediately call the doctor.

Hydrogen bromide gas: On inhalation, first aid is to inhale fresh air, lie down and rest.

Hydrogen chloride gas: Same as hydrogen bromide gas.

Hydrogen cyanide: It is a flammable, fire hazard gas and causes headache, nausea, unconsciousness and death.

Hydrogen peroxide: It is a powerful oxidising agent and concentrated solution can result in explosion.

Hydrogen sulphide gas: If inhaled, first aid is to inhale fresh air. Give artificial respiration, if required.

Hydroidic acid: Same as hydrobromic acid.

Hydroxylamine: It is highly hygroscopic and unstable. It results in dizziness, headache, jaundice, vomiting and difficult breathing.

Iodine: It causes cough, dizziness, headache and difficult breathing.

Mercury: It is a heavy liquid, which on inhalation causes cough, fever, abdominal cramps and inflammation of mouth, while on swallowing causes nausea, bloody diarrhoea and burning in the mouth and throat. Long exposure causes death.

Nitrobenzene: It can be absorbed in fatal doses when comes in contact with skin.

Peracids: e.g. Peracetic acid, trifluoroacetic acid. They are powerful oxidising agents and their concentrated solutions are highly explosive.

Peroxides: All peroxides are powerful oxidising agent and extremely explosive, especially when heated. Their mixtures with oxidisable materials are also highly explosive.

Phenol: It is a corrosive liquid and should be handled with gloves in hands. When comes in contact with skin, it can be absorbed in fatal doses. It causes cough, vomiting and burns in mouth and pharynx.

Phenylhydrazine: It can also be absorbed in fatal doses when comes in contact to skin.

Phosgene: It is a extremely toxic, non-combustible liquid. It causes cough, headache, vomiting, rapid respiration and pain in the upper abdomen. Inhalation for long causes death.

Phosphoric acid: When spills, wash with lots of water followed by sodium bicarbonate solution.

Phosphorus pentachloride: It is a flammable powder, which causes bronchitis, irritation of eyes and inflammation of kidney.

Potassium cyanide: It is extremely toxic and fatal in small amounts. Even a cut in the skin can ingest the fatal doses. If spills, it should be cleaned carefully and immediately disposed of. On acidification, its solution gives off hydrogen cyanide, which is a deadly gas.

Potassium hydroxide: It is a corrosive solution and if spills, then first wash with plenty of water and finally with dilute bicarbonate solution.

Potassium metal: It reacts violently with water to ignite hydrogen gas. It should be stored in the absence of oxygen otherwise, it forms extremely dangerous and explosive peroxides, and such materials should not be used but disposed of. Potassium metal is usually stored in xylene. For oxide-free metal, the oxide layer is cut off with a knife and scrap of potassium should be disposed of by adding excess of tert-butanol to small pieces.

Pyridine: It is a flammable liquid with nauseating odour and causes conjunctivitis, headache, itching, vomiting and abdominal pain.

Reactive halides: Aluminium chloride, boron trichloride, phosphorus tribromide, trichloride and pentachloride and titanium tetrachloride are highly reactive, especially towards water. All of them are corrosive and irritant.

Sodamide: It is irritant and corrosive in finely divided form. It reacts explosively with water. Old samples decompose explosively and so should not be powdered for use but should be destroyed.

Sodium cyanides: Same as potassium cyanide.

Sodium hydroxide: Same as potassium hydroxide.

Sodium metal: Same as potassium metal, the only difference is that in sodium metal, its scraps should be disposed of by adding excess of methanol.

Styrene: It causes conjunctivitis, nausea, weakness and loss of appetite.

Sulphuric acid: It is highly corrosive and if spills, wash with excess of water followed by sodium bicarbonate solution.

Thionyl chloride: It is a highly corrosive, volatile liquid, which causes conjunctivitis, pneumonia and skin inflammation. If spills, first wash with plenty of water and then finally with dilute ammonia solution.

Toluene: It is a flammable liquid, which causes nausea, weakness and inflammation on skin.

Following is the list of several compounds depending upon their toxicological properties:

1. *Highly toxic solids:* These solids are harmful in even small quantities and may cause rapid serious illness or even death and, therefore, great care should be taken to prevent their entrance in body by inhalation of dust, absorption through skin or ingestion.

   e.g. Inorganic cyanides, arsenic and mercury compound, oxalic acid, vanadium pentoxide.
2. *Highly toxic liquids:* These liquids are highly irritant, mainly to the eyes and respiratory tract. Absorption of their vapours through skin is also very harmful and results in chronic effects on long-term exposure to small amounts.

   e.g. carbon disulphide, bromomethane, benzene, tetrachloro and tetrabromoethane, hydrofluoric acid, dimethyl sulphate, allyl alcohol, crotonaldehyde.
3. *Highly toxic gases:* These gases are highly toxic and therefore, their contact with skin must be avoided. And the best precaution is to carry out all the experiments involving them in an effective fuming cupboard.

   e.g. carbon monoxide, cyanogen, hydrogen cyanide, hydrogen sulphide, nitrogen dioxide, hydrogen fluoride, boron trifluoride.
4. *Carcinogens:* They are cancer-causing agents, and therefore, great care should be taken to prevent the inhalation of their vapours and absorption through skin. While handling them, gloves should be worn and their bottles should be kept in fuming cupboard with proper labelling of the warning.

e.g. α- & β- naphthylamine, diazomethane, dimethyl sulphate, methyl iodide, thiourea, thioacetamide, benzidine and certain polynuclear hydrocarbons.

Table 1.2 gives the carcinogens and suspected carcinogens in a laboratory.

Table 1.2 List of carcinogens & suspected carcinogens.

Carcinogenic in humans	*Carcinogenic in animals*	*Suspected to be Carcinogenic*
Arsenic (III) & (IV) Oxides	Acrylonitrile	Acetamide
Arsenic acids & salts	Diazomethane	Allyl chloride
Asbestos (as dust)	1,2-Dibromoethane	Benzalchloride
Benzene	Dimethyl sulphate	Benzyl chloride
Benizidine & salts	Aziridine	Chlorinated biphenyls
Coal Tar	Hexamethylphosphoric triamide (HMPA)	Chloroform
2-Naphthylamine	Hydrazine	Chromium (VI) oxide
Vinyl chloride	Methyl iodide	1,2-Dichloroethane
	2-Nitronaphthalene	1,4-Dioxine
	2-Nitropropane	Phenylhydrazine
		O-Tolidine
		O-Toluidine
		1,1,2-Trichloroethane
		Vinylidene chloride

1.2 FIRST-AID

A clearly labelled, first aid box should be kept in a readily approachable place in the laboratory. It should contain the following items, all of which should be clearly labelled:

Several sizes bandages, lint, cotton-wool, adhesive plaster, and a sling;

Scissors, safety pins, delicate forceps, needles and thread;

Fine glass dropper and two glass eye-bath;

Vaseline, castor oil, olive oil, Sal volatile, Boric acid powder, chloramine-T powder, sodium bicarbonate powder, sulphapyridine powder;

Acriflavine jelly (e.g. burnol), tannic acid jelly (e.g. Tannafax);

Bottles of the following:

Savlon antiseptic, 1% acetic acid, 1% boric acid, 1% sodium bicarbonate solution, saturated sodium bicarbonate solution, rectified spirit, glycerine, light petroleum (80-100°C boiling fraction).

A fire-proof blanket should be stored in a labelled container near the first aid box.

1.3 APPARATUS USED IN THE LABORATORY

In an organic chemistry laboratory, a wide range of glass apparatus are used for carrying out the synthesis, separation and purification of organic compounds. The students should be familiar with the use and handling of the apparatus. This section deals with the general laboratory apparatuses employed in organic chemistry laboratory.

In general, the small scale apparatus is used for organic preparations. Various advantages of carrying out the organic preparative works on small scale are:

(a) The cost of small scale apparatus is low and also the expenditure due to breakages etc. is less than the corresponding macro apparatus.

(b) As the laboratory budget for chemicals is very low and therefore, the costly chemicals can be used in classes of large number of students because of requirement of small amounts of chemicals.

(c) Since the time required for the completion for the preparation on small scale is less than that of the macro scale, more preparations can be carried out and hence more experience for work in a limited period of time.

(d) The chances of accidents due to fire, explosions etc. is greatly reduced for small scale preparations.

(e) The student can now work on the small scale preparations with ease.

(f) The student can store these glass apparatuses in a comparatively smaller locker.

(g) Also, less bench space is required by each student.

A brief discussion of the general types of glass apparatus is given here:

1.3.1 Flasks

Various types of flasks are shown in Fig. 1.1(a-g)

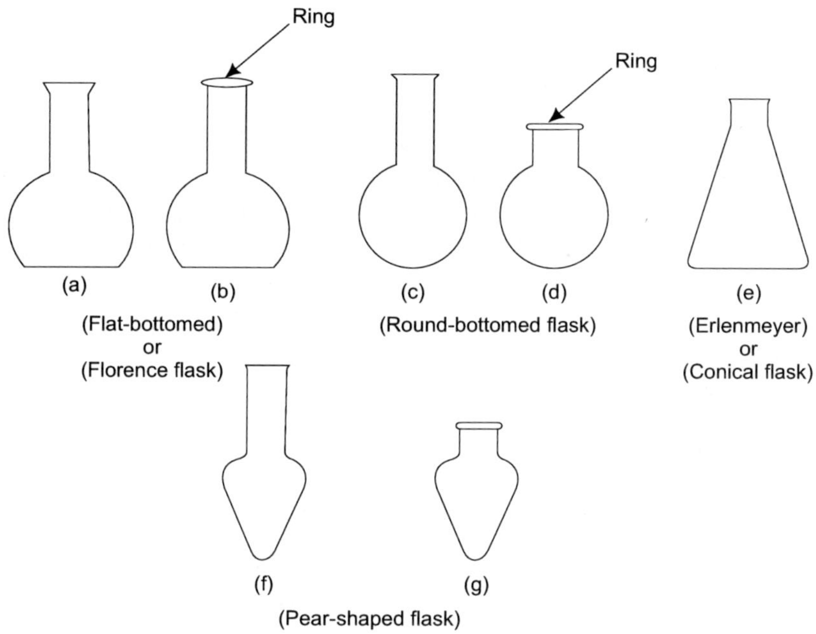

Fig. 1.1 (a-g) Various types of flasks

Type (a) is a flat bottomed flask, also called florence flask of capacities 50 ml, 100 ml, 150 ml, 200 ml and 250 ml with 19 mm external diameter of the neck.

Type (b) is same as (a) except includes a ring neck, which provides the mechanical strength to it.

Type (c) Round bottomed flask having capacities 25 ml, 50 ml, 100 ml, 150 ml, 200 ml and 250 ml with 19 mm external diameter of the neck.

Type (d) Round bottomed flask with a ring neck, also called bolt head flask. It is widely used and is availables in 25 ml, 50 ml, 100 ml, 200 ml and 250 ml capacities.

Type (e) includes Erlenmeyer or conical flask having narrow and wide mouth designs with or without graduations and the sizes include 5 ml, 10 ml, 25 ml, 50 ml, 150 ml, 200 ml and 250 ml.

Type (f) Pear-shaped flask having capacities of 5 ml, 10 ml, 25 ml and 50 ml.

Type (g) Pear-shaped flask with ring neck having capacities of 5 ml to 50 ml.

These pear-shaped flasks are used preferably for small volumes of liquids than the other flasks as losses due to drainage are reduced.

1.3.2 Beakers

The types of beakers are shown in Fig. 1.2 (a-b).

Type (a) is called griffin form with spout, and is widely used. The sizes include 5 ml, 10 ml, 15 ml, 20 ml, 30 ml, 50 ml, 100 ml, 150 ml, 200 ml, 250 ml, 400 ml and 600 ml.

Type (b) is tall form without spout and is used occasionally in preparative work.

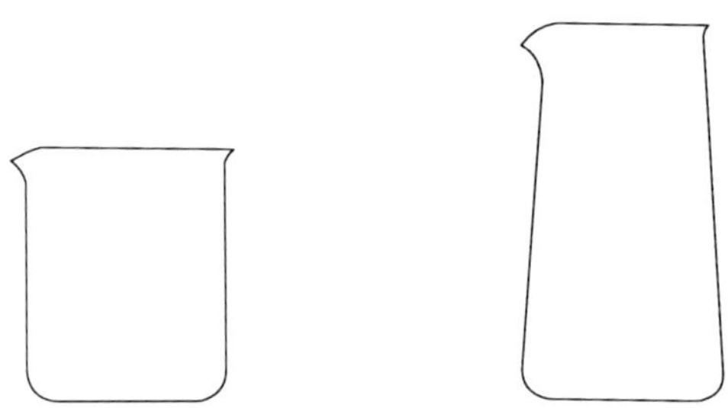

(a)
(Griffin form with spout)

(b)
(Tall form without spout or conical beaker)

Fig. 1.2 (a-b) Types of beaker

1.3.3 Condensers

For small scale, various types of condensers are shown in Fig. 1.3 (a-c).

Type (a) is a condenser jacket for a Liebig-type condenser.

Type (b) is an all-glass Liebig-type condenser. It consists of an inner glass tube, which is surrounded by a glass jacket for the circulation of water. It is available in three useful sizes with jackets of 3", 4" and 5" lengths.

Type (c) is a double surface condenser of the Davies type. Here the jacket is having 4" length. It is more efficient than the type (b) of the same jacket length.

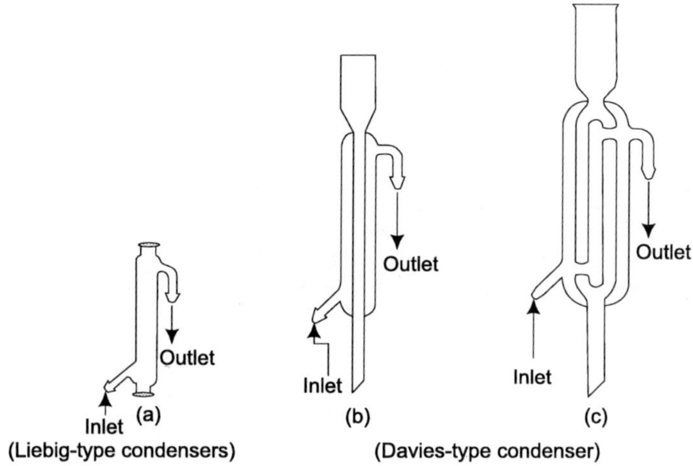

Fig. 1.3 (a-c) Types of condenser

1.3.4 Distilling Flask

Various distilling flasks employed for small scale work are shown in Fig. 1.4 (a-f)

Type (a): is a distilling flask with a pear-shaped bulb and a long side arm. Condenser is attached to its side arm as shown in Fig. 1.5. Its common sizes are 5 ml, 10 ml, 25 ml and 50 ml.

Type (b): is an ordinary distilling flask having spherical bulb with a long side arm. The various common sizes of this type are 10 ml, 25 ml and 50 ml.

Type (c): is a pear-shaped flask with elongated side arm. Various common sizes are 5 ml, 10 ml, 25 ml and 50 ml.

Type (d): is a distilling flask with spherical bulb and elongated side arm. The sizes include 10 ml, 25 ml and 50 ml.

Type (e): is a distilling flask having a pear-shaped bulb with a fractionating side, which can be charged with a suitable packing. The sizes include 25 ml and 50 ml and the effective length of packing is 12.5-15 cm.

Type (f): is similar to the type (e), only difference is that it has a spherical bulb. In this type, the common sizes are 100 ml, 175 ml and 250 ml with 15 cm effective length of packing.

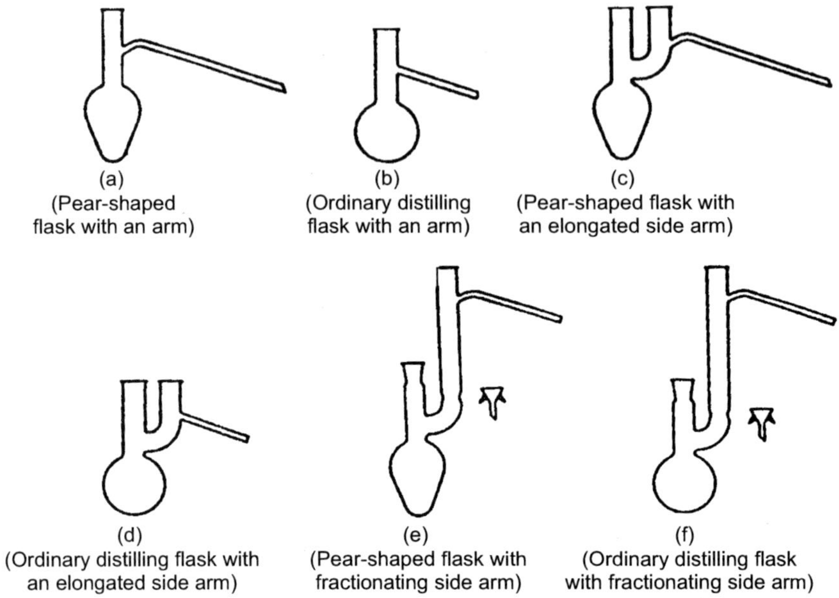

Fig. 1.4 (a-f) Types of distilling flask

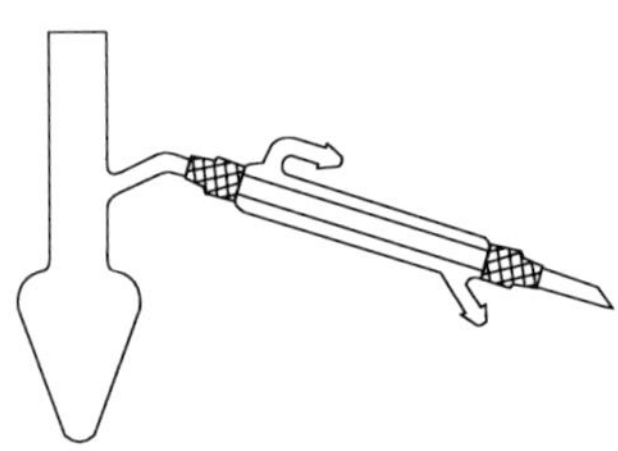

Fig. 1.5 Attachment of the condenser to the side arm of type (a) flask

1.3.5 Multinecked Flasks

Various types of multi-necked flasks are shown in Fig. 1.6 (a-c). In such flasks, the central aperture is usually larger than the side apertures.

Type (a): It is a two-necked flask having a central and a side neck.

Type (b): It is a three-necked flask having a central and two side necks. For 250 ml capacity, the diameter for central aperture is 30 mm while for side apertures is 17 mm, and for smaller flasks, say 150 ml, the diameter for central aperture is 28 mm while for side apertures is 17 mm.

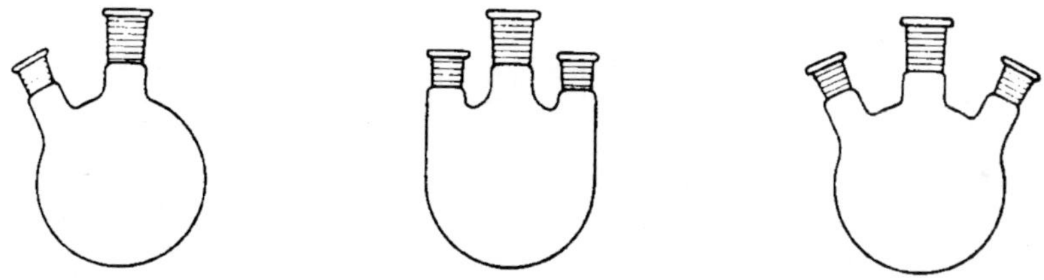

(a) (Two-necked flask) (b) (Three-necked flask) (c) (Three-necked flask with inclined side necks)

Fig. 1.6 (a-c) Types of multinecked flask

Type (c): It is similar to type (b) except here the side-apertures are inclined at an angle which make it convenient for fitting the apparatus into these side necks.

These flasks are widely used for preparation in organic chemistry. The central aperture is fitted with a mechanical stirrer while the two side-apertures are attached with a reflux condenser and a dropping funnel.

1.3.6 Funnels

Various types of funnels are shown in Fig. 1.7 (a-f)

Type (a): It is an ordinary 60° funnel, having 50 mm diameter.

Type (b): It is the type (a) funnel with short stem and is very useful in many cases.

Type (c): It is a globular separatory funnel. The common sizes include 10 ml, 25 ml and 50 ml.

Type (d): It is a pear-shaped separatory funnel. It is more preferable to type (c). It should be light weight so that it can be inserted into the multinecked flask without any clamping.

Type (e): It is a cylindrical shaped graduated separatory funnel.

Type (f): It is also called as dropping funnel. It is similar to that of type (c) except that here, the liquid can be added drop-wise due to the presence of a device exactly below the tap. It has the advantage that the rate of flow of liquid can be visualized even when the stem of funnel is dipped in a liquid.

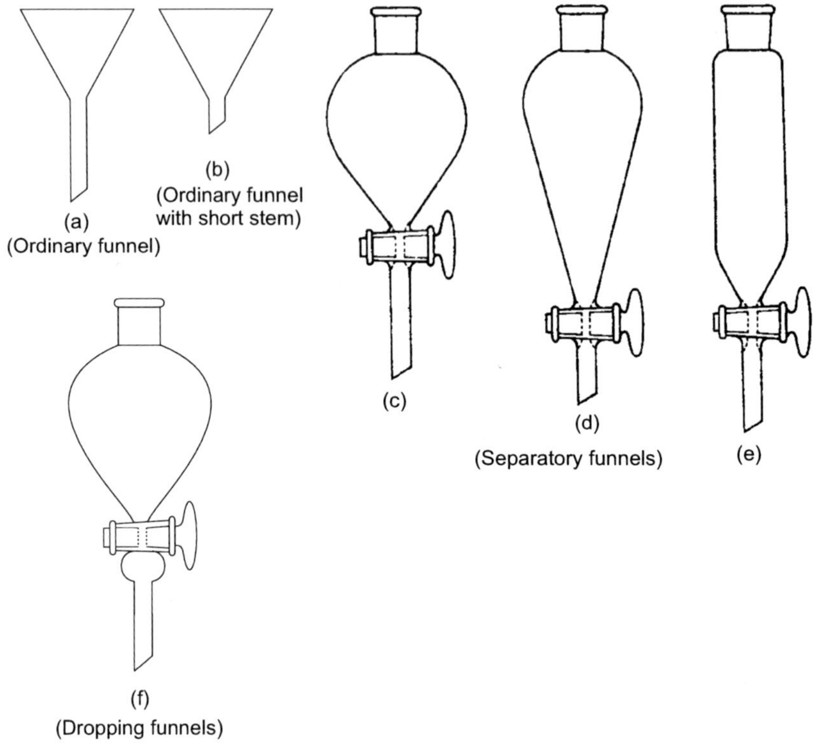

Fig. 1.7 (a-f) Types of funnel

1.3.7 Hot Water Funnels

Various types of hot water funnels are shown in Fig. 1.8 (a-c).

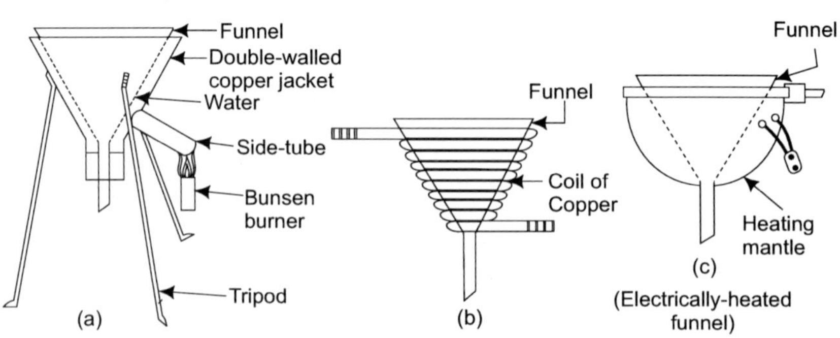

Fig. 1.8 (a-c) Types of hot water funnel

Type (a): It consists of a double walled copper jacket (to keep the funnel into it) on tripod. Water is filled in the space between the walls and can be heated to the required temperature (less than 100°C) by using a Bunsen burner.

Type (b): It is a coil of copper (10 mm diameter) in the form of 60° cone to keep the funnel into it. Through it, hot water or steam is circulated. It is an efficient method for the filtration of inflammable liquids. If the cold water is circulated through the coil, it may be used in cold filtration.

Type (c): It is an electrically heated funnel. Here electrically heating mantle is used as a house for the funnel to be heated.

1.3.8 Funnels for Filtration by Suction

These types of funnels are shown in Fig. 1.9 (a-f).

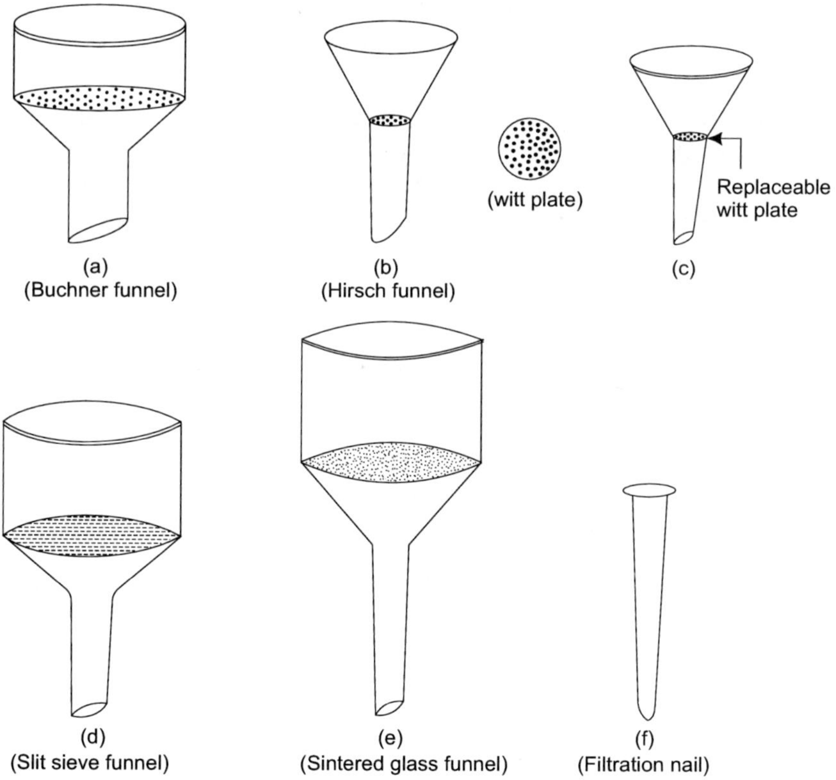

Fig. 1.9 (a-f) Types of filtration funnels

Type (a): It is a Buchner funnel. It is made up of porcelain and has a perforated porcelain plate of different sizes. The most commonly used sizes are having the plate diameters 42 mm & 56 mm.

Type (b): It is a Hirsch funnel having sloped sides. It is a modification of Buchner funnel and is used for smaller amounts of solid product. The smallest size can hold a filter paper of 3-4 mm in diameter. It is also available in various sizes, but the most commonly used sizes are having the plate diameters of 27 mm and 47 mm, and the top diameters of 65 mm and 78 mm respectively.

Type (c): It is the modification of Hirsch funnel. It consists of an ordinary glass funnel with a perforated porcelain plate called as witt plate. The diameter of the witt plate is 1-4 cm. The advantage of this funnel over the above two funnels is that as it is replaceable and so the apparatus can be cleaned properly while in the type (a) & (b) funnels lower side of the perforated plate cannot be examined.

Type (d): It is slet sieve funnel. It is an all-glass funnel and therefore more advantageous than the porcelain funnels which are opaque. The most useful size is the 60 mm diameter disc.

Type (e): It is a sintered glass funnel with a number of porosities i.e. coarse, medium and fine. The most convenient sizes of the discs are 30 mm, 40 mm, 50 mm and 65 mm diameter.

Type (f): It is a filtration nail and is very useful for small amount of precipitates. It consists of a thin glass rod flattened at one end. It can be formed by heating one end of a thin short glass rod in the blowpipe flame and then pressing on the asbestos board vertically.

Fig. 1.10 (a-c) illustrates the assemblies for filtration by suction of above types of funnel.

Type (a), (d) & (e) funnels are fitted in a filtration flask with the help of a rubber stopper as shown in Fig. 1.10 (a). The filtration flask is then attached to a water suction pump by means of a pressure tubing via a suitable trap.

Type (b) & (c) and for nails, the funnels are fitted on a filtration tube with the help of a rubber stopper as shown in Fig. 1.10 (b) & (c)

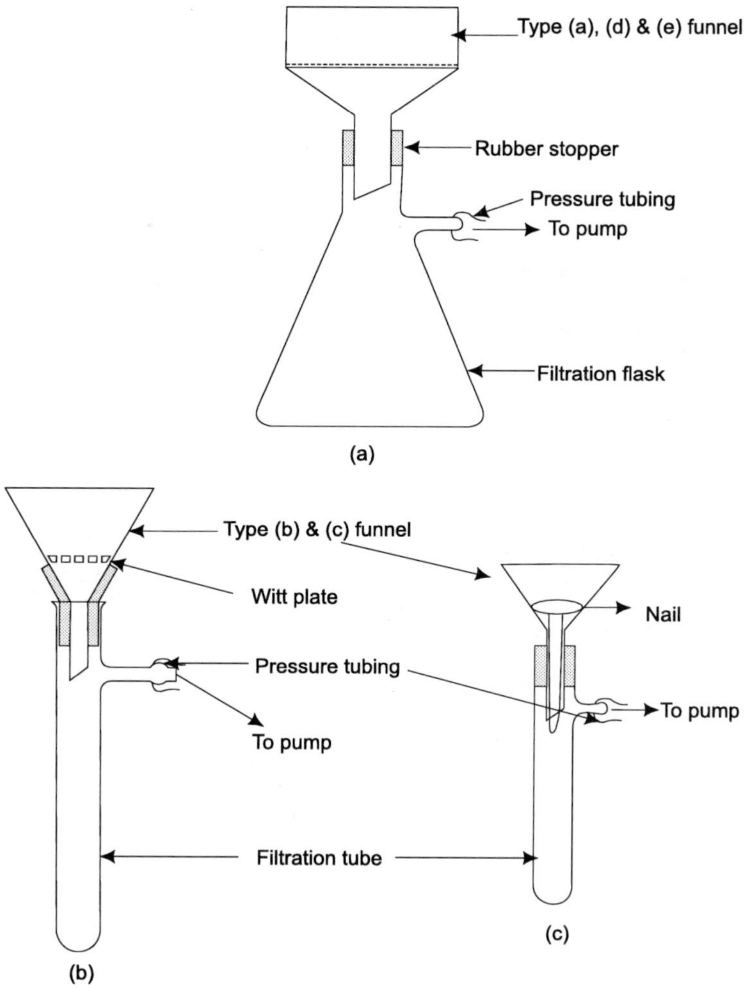

Fig. 1.10 (a-c) Assemblies for suction filtration by various funnels

respectively and the side-arm of the tube is then attached to a water suction pump with a pressure tubing via a suitable trap.

1.3.9 Spatulas

Commercially available spatulas are nickel and stainless steel spatulas. Two types of these spatulas are shown in Fig. 1.11 (a-b).

Type (a): It is a nickel spatula with turned-up end to handle the small amounts of material. It is available in various sizes, the most common is 100 mm length and 4 mm width of the blades.

Type (b): It is a nickel double spoon ended spatula.

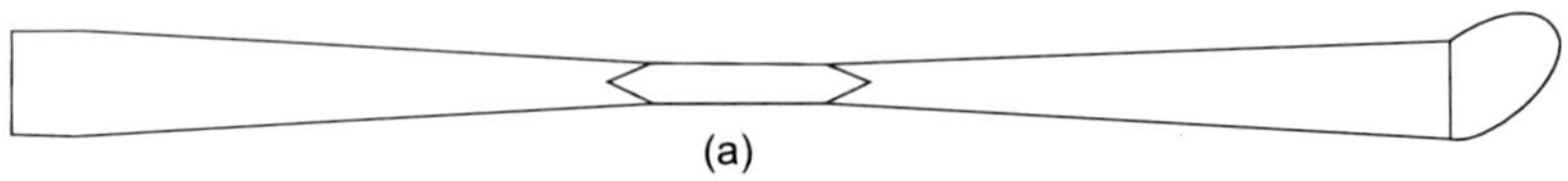

(a)
(Nickel spatula with turned up end)

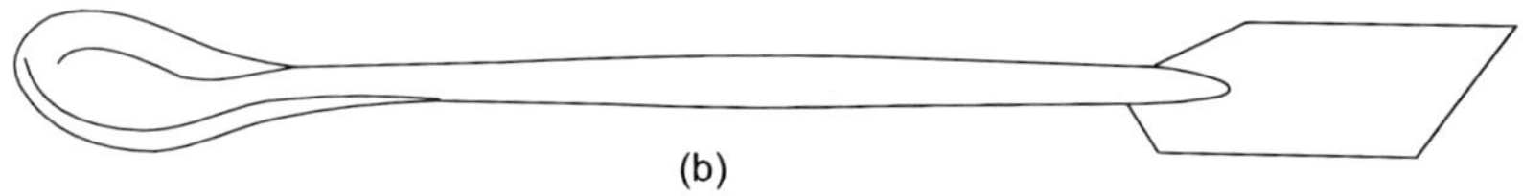

(b)
(Nickel spatula with double spoon)

Fig. 1.11 (a-b) Types of spatula

1.3.10 Adapters

During assembly of the apparatus, various adapters are required. They are of following types:

1.3.10.1 Joint adapters

They are required if the joints of various parts are not consistent. They are further classified into two types, which are illustrated in Fig. 1.12 (a-b).

(a) (Reduction Adapter) (b) (Expansion Adapter)

Fig. 1.12 (a-b) Types of joint adapter

Type (a): It is reduction adapter as shown in Fig. 1.12 (a).

Type (b): It is an expansion adapter [Fig. 1.12(b)].

Various combinations of these adapters are available. Their use should be avoided to form a well-designed assembly of the apparatus.

1.3.10.2 Multiple adapters

For multiple entries in a reaction system, one way is to use multi-necked flasks while other option is to use multiple adapters. Various multiple adapters are shown in Fig. 1.13 (a-c).

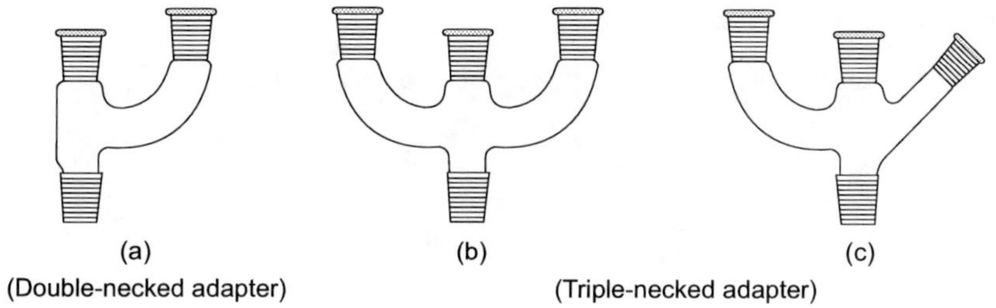

Fig. 1.13 (a-c) Types of multiple adapter

Type (a): It is a double-necked adapter as shown in Fig. 1.13 (a).

Type (b): It is a triple-necked adapter [Fig. 1.13(b)].

Type (c): A triple-necked adapter as shown in Fig. 1.13(c).

1.3.10.3 Receiver adapters

They are also called as connectors and are used to connect the condenser to the receiver. Their types are illustrated in Fig. 1.14(a-c).

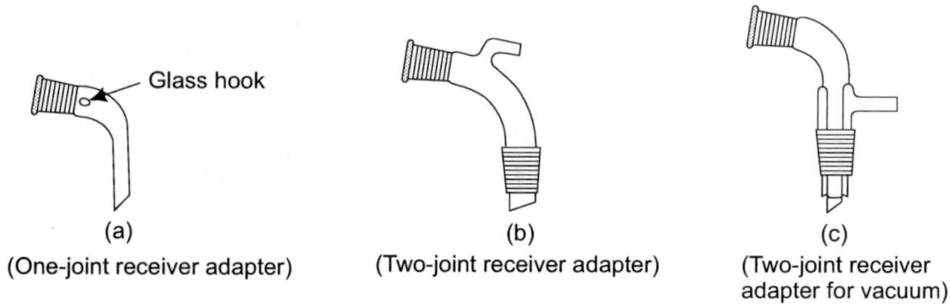

Fig. 1.14 (a-c) Types of receiver adapter

Type (a): It is the simplest receiver adapter with glass hooks to hold it by the condenser with the help of a rubber band from these hooks.

Type (b): It is a two-glass joint receiver adapter, better than type (a).

Type (c): It is a two-glass joint receiver adapter with a side tube as shown in Fig. 1.14 (c). It is more useful than the above two types because this side tube can be connected to a vacuum pump when required and so distillation under reduced pressure can be carried out by this type of adapter.

1.3.10.4 Take off adapters

They are cone/rubber tubing adapters and are shown in Fig. 1.15 (a-b).

Fig. 1.15 (a-b) Types of take off adapter

They are widely used in preparations in organic chemistry e.g. removal of small amounts of solvents rapidly (without recovering them).

1.3.11 Distillation Heads

They are also called as still-heads and are illustrated in Fig. 1.16 (a-c).

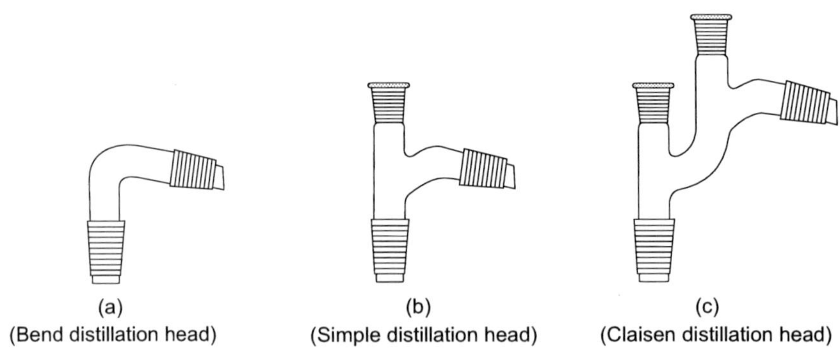

Fig. 1.16 (a-c) Types of distillation head

Type (a): It is a bend head, which is generally used for the distillation involving only the removal of solvent.

Type (b): It is a simple distillation head. It is fitted into the distillation flask. A thermometer dipped in mercury of mercury jacket (shown in Fig. 1.17) is placed on the upper socket and a condenser on the side-way socket.

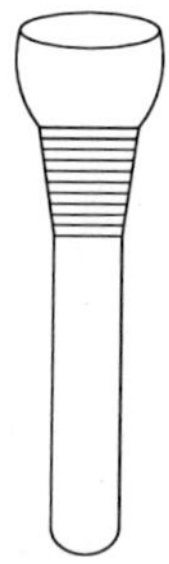

Fig. 1.17 Mercury Jacket

Type (c): It is a claisen distillation head. Here the right hand socket is used for thermometer and left hand socket for capillary tubes used in distillation under vacuum. The side-way socket is for condenser.

1.3.12 Stoppers

Stoppers of ground glass joints are available in all standard sizes, one of which is shown in Fig. 1.18.

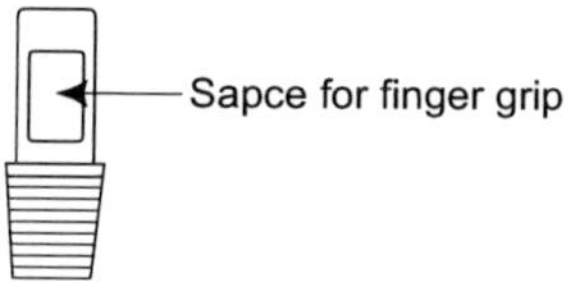

Fig. 1.18 Stopper

The preferable stopper is the one having the flat head so that can be stood when not required and thus prevents the sticking of impurities from the surface on the joint.

1.3.13 Guard Tube

Calcium chloride guard tube is shown in Fig. 1.19.

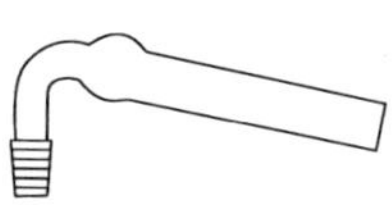

Fig. 1.19 Guard tube

It is filled with anhydrous calcium chloride keeping the cotton plug at both the ends as illustrated in Fig. 1.20.

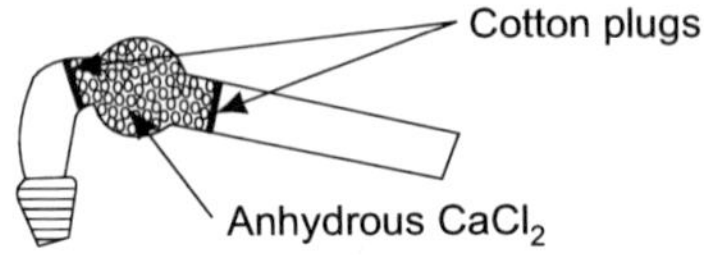

Fig. 1.20 Guard tube filled with calcium chloride

It is widely used to protect the substances or assemblies of apparatus from moisture by attaching it to the apparatus.

1.3.14 Desiccator

Desiccators are of two types: an ordinary desiccator and a vacuum desiccator, which are described below:

1.3.14.1 Ordinary desiccators

These are the desiccators used at atmospheric pressure, and are available in dry-seal and ground flange range and have limited use. It is illustrated in Fig. 1.21.

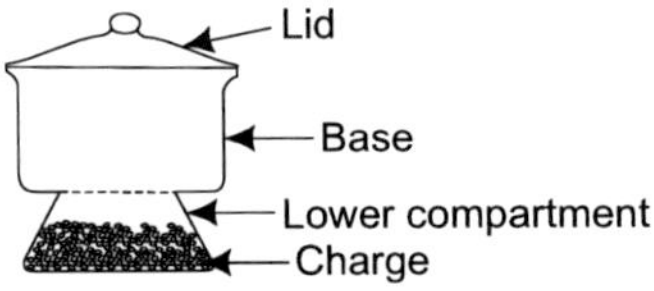

Fig. 1.21 Ordinary desiccator

The nature of the substance (charge) to be filled in the lower compartment (shown in Fig. 1.21) depends upon the nature of the drying process i.e. whether water or organic solvents is to be removed and whether acidic or basic vapours are evolved in this process.

For example, suitable charges to absorb

(1) water vapours are anhydrous calcium chloride, concentrated sulphuric acid or silica gel,
(2) methanol and ethanol vapours are anhydrous calcium chloride and silica gel,

(3) vapours of benzene, toluene, light petroleum, chloroform, carbontetrachloride and diethyl ether are freshly cut shavings of paraffin wax,

(4) vapours moist with acetic acid or hydrochloric acid, dual charge is required, one for absorbing moisture i.e. silica gel or concentrated sulphuric acid and another for acids i.e. flake of sodium hydroxide.,

(5) moisture from the samples to be used in anhydrous reaction conditions is phosphoric oxide.

1.3.14.2 Vacuum desiccator

It is illustrated in Fig. 1.22

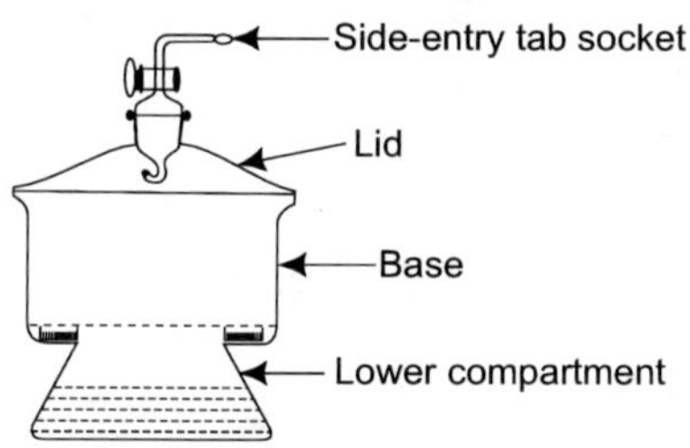

Fig. 1.22 Vacuum dessicator

Vacuum desiccator is used widely to dry the moist with water or organic solvents. Here, the side-entry tap socket is fitted into the lid. In desiccator, air-inlet is shaped as a hook (as shown in Fig. 1.23) to prevent the scattering of the sample during the application and removal of vacuum.

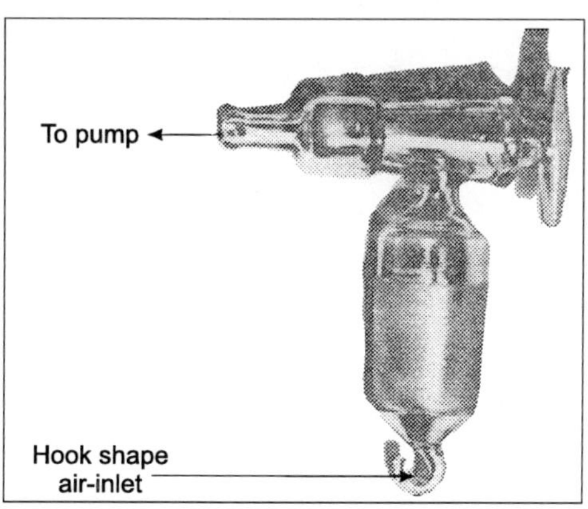

Fig. 1.23 Side-entry tap socket

The lubricant should be applied to the joint between the lid and the base, before the vacuum is created in the desiccator.

Another vacuum desiccator, which is recently introduced, is with dry-seal joint, illustrated in Fig. 1.24.

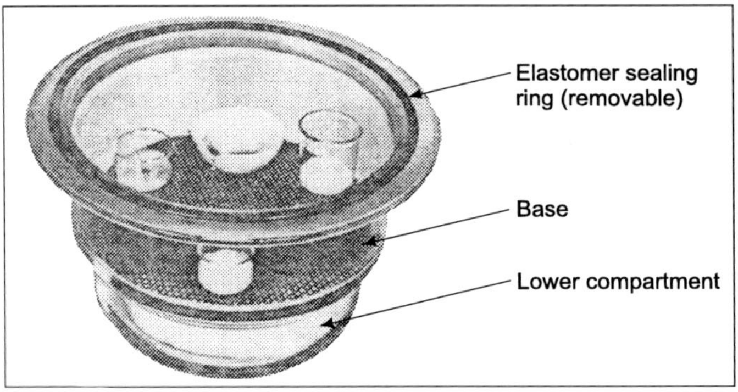

Fig. 1.24 Desiccator-base with dry-seal joint

It has a removable elastomer sealing ring as shown in Fig. 1.24, which gets flattened by the lid, when desiccator is evacuated. When the vacuum is released, the lid is lifted off easily and, therefore, lubrication is not required.

1.3.15 Advantages of Interchangeable Ground Glass Joints

The advantages of apparatus consisting interchangeable ground glass joints over the apparatus which require corks or rubber stoppers for assembly are as follows:

(a) It is easy to handle the corrosive solids and liquids, e.g. concentrated acids, bromine, phosphoruspentachloride etc.

(b) No contamination of impurities from apparatus (i.e. cork or rubber stopper) into the product.

(c) As no corks/rubber stoppers are required and therefore no wastage of time in the boring and fitting of corks.

(d) All joints fit well as they are made accurately standard.

(e) Apparatus assembly is aligned accurately. Generally, incorrect bored corks result in inaccurate alignment.

(f) Assembling of the apparatus is very fast.

(g) Since all joints of the same size are interchangeable, it is easy to replace the broken part without any fear of improper fit.

(h) Most of the common operations of organic chemistry can be carried out by using a few units.

(i) The danger of violent reactions are decreased as the vapours now have wider passage to move.

While using ground glass joint apparatus, following precautions should be taken into account:

(a) They must be free from any impurity e.g. dust etc.

(b) Very small quantity of vaseline or grease should be applied around the outer part of each of the glass joint to prevent sticking of the apparatus.

(c) For distillation under reduced pressure, all glass joints must be lubricated with vaseline or apiezon grease or a silicone grease.

(d) Apparatus should be parted after use when it is still warm to avoid sticking of the apparatus, especially when alkalis are used. Alkalis have the tendency to stick the joints and, therefore, while using them, the joints should be properly lubricated and the apparatus should be dismantled immediately after use.

(e) All the components in the assembly should be properly clamped with the clamps whose jaws are lined with any soft material e.g. cork, rubber.

(f) After dismantling joints must be cleaned carefully.

Even after taking all the precautions, if the sticking of the apparatus results, then the following steps should be carried out to freeze it:

(1) Hold the joint in the vertical position and apply glycerine to the upper surface. This will help in the seizing of the apparatus as the glycerine will slowly and slowly penetrate into the joint.

(2) Draw a stream of hot air on the outer surface of the joint for a few seconds with the help of a blower and then apply a twisting action to move them apart gently. Even tapping of the seized joint on the edge of wooden bench gently may result in the separation of these joints.

(3) Heat the joint into a small Bunsen flame for a few seconds and pull them apart gently. Heating will result in the expansion of the socket glass to make it separate from the other joint.

1.4 APPARATUS USED FOR DIFFERENT TYPES OF REACTIONS

In this section, the various assemblies to carry out different reactions in a organic laboratory are discussed. The apparatuses used to perform a reaction can be as simple as a test tube or boiling tube or flask. However, generally the assembly of apparatuses is made which consist of a flask and a condenser along with a drying tube to prevent the moisture, a separatory funnel to add the components into the flask, a trap to absorb the evolved gases etc.

1.4.1 A Reaction Under Reflux

Fig. 1.25 illustrates the assembly used for reactions under reflux. In this assembly, a condenser is used to bring down the vapours to the flask to ensure no loss in the contents of flask during the refluxing. Boiling stones (pumice stones) should always be added in the reaction mixture before starting the heating to ensure smooth refluxing.

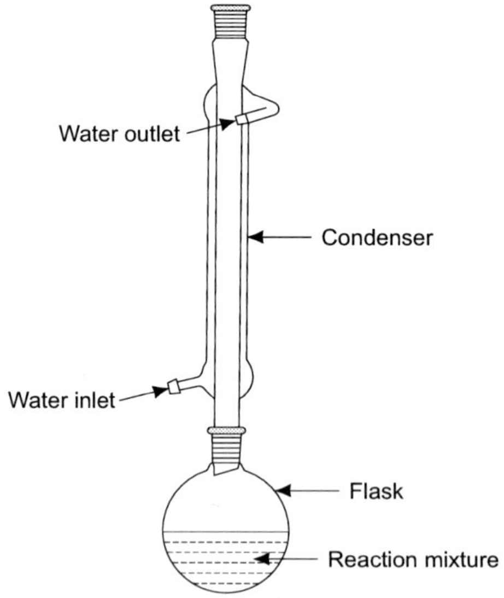

Fig. 1.25 Apparatus for a reaction under reflux

For reactions in anhydrous conditions, a guard tube charged with anhydrous calcium chloride is fitted at the end of the condenser to exclude the moisture. Such an assembly is shown in Fig. 1.26.

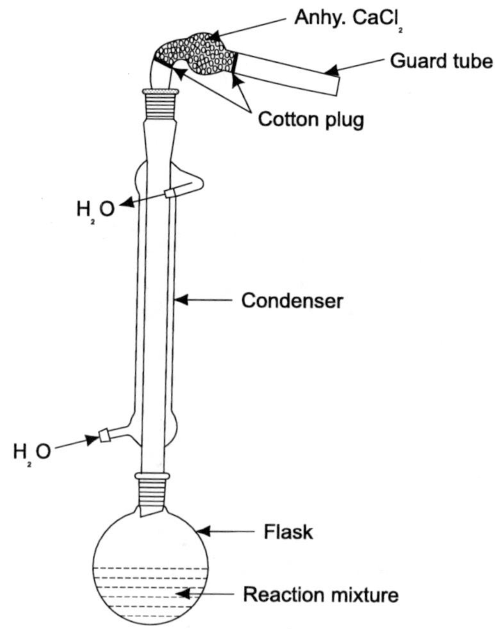

Fig. 1.26 Apparatus for a reaction under reflux with a guard tube

1.4.2 A Reaction Under Reflux with Addition of a Liquid

Fig. 1.27 illustrates assembly used for such reactions. Here by using an adapter one-necked flask is converted into two-necked flask. Alternatively, flask and double-necked adapter may be replaced by a two-necked flask. To ensure thorough mixing manual shaking is required at intervals. If the dry conditions are required in such reactions a calcium chloride guard tube is to be fitted at the end of the condenser in Fig. 1.27.

1.4.3 A Reaction Under Reflux with Addition of a Liquid with Stirring

Apparatus required for such type of reaction is shown in Fig. 1.28.

For homogeneous reaction mixture, above two types of apparatus assemblies can be used while with heterogeneous reaction mixtures (either due to immiscible liquids or suspended solid components) stirring is necessary. This stirrer may be replaced by a magnetic stirrer having an electric hot place.

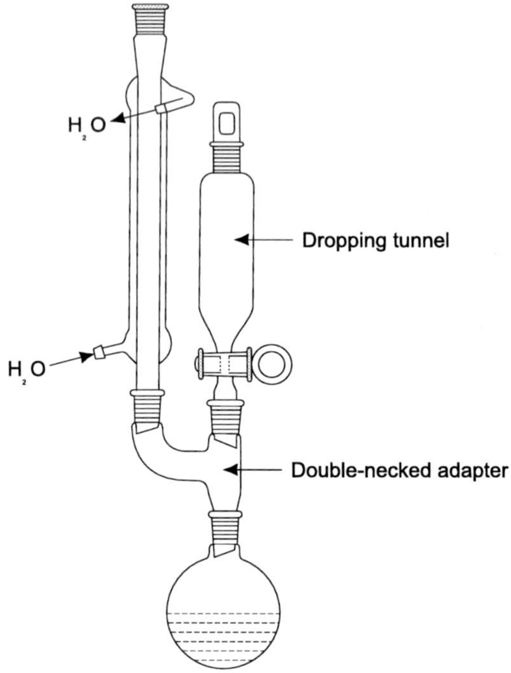

Fig. 1.27 Apparatus for a reaction under reflux with addition of a liquid

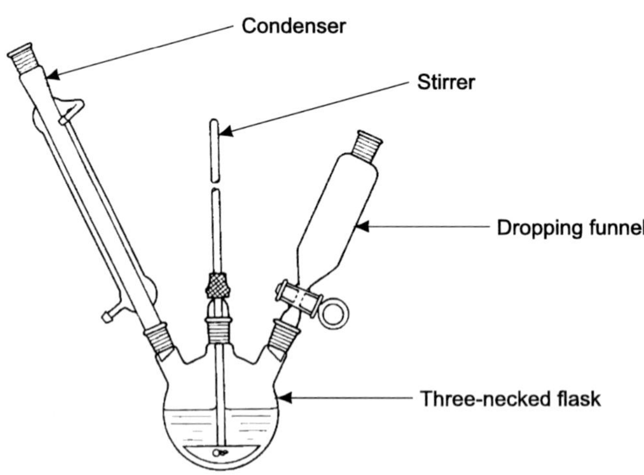

Fig. 1.28 Apparatus for a reaction under reflux with addition of a liquid with stirring

For anhydrous conditions, a guard tube charged with anhydrous calcium chloride is to be placed at the end of the condenser in Fig. 1.28.

1.4.4 A Reaction Under Reflux with Addition of a Liquid to a Stirred Solution with Controlled Temperature

Fig. 1.29 illustrates the assembly used for such reactions.

If dry atmosphere is required, guard tube is to be fitted at the end of the condenser as shown in Fig. 1.29, otherwise not required. In this apparatus, a three-necked flask is converted into four-necked flask by using an adapter.

For the small quantities, mechanical stirrer (as shown in Fig. 1.29) may be replaced by a magnetic stirrer.

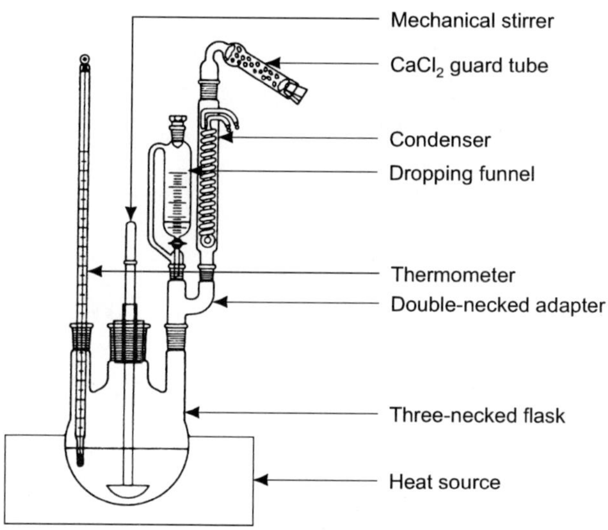

Fig. 1.29 Apparatus for a reaction under reflux with a addition of a liquid to a stirred solution with controlled temperature

1.4.5 A reaction under Reflux with Addition of a Solid Reagent to a Stirred Solution in a Dry Atmosphere

Fig. 1.30 (a-c) illustrates the assembly used for this type of reaction.

In Fig. 1.30 (a), the solid is added using a dropping funnel having an adapter with o-ring. The solid is added into the reaction mixture by raising the glass rod while the flow of solid is stopped by moving down the rod to

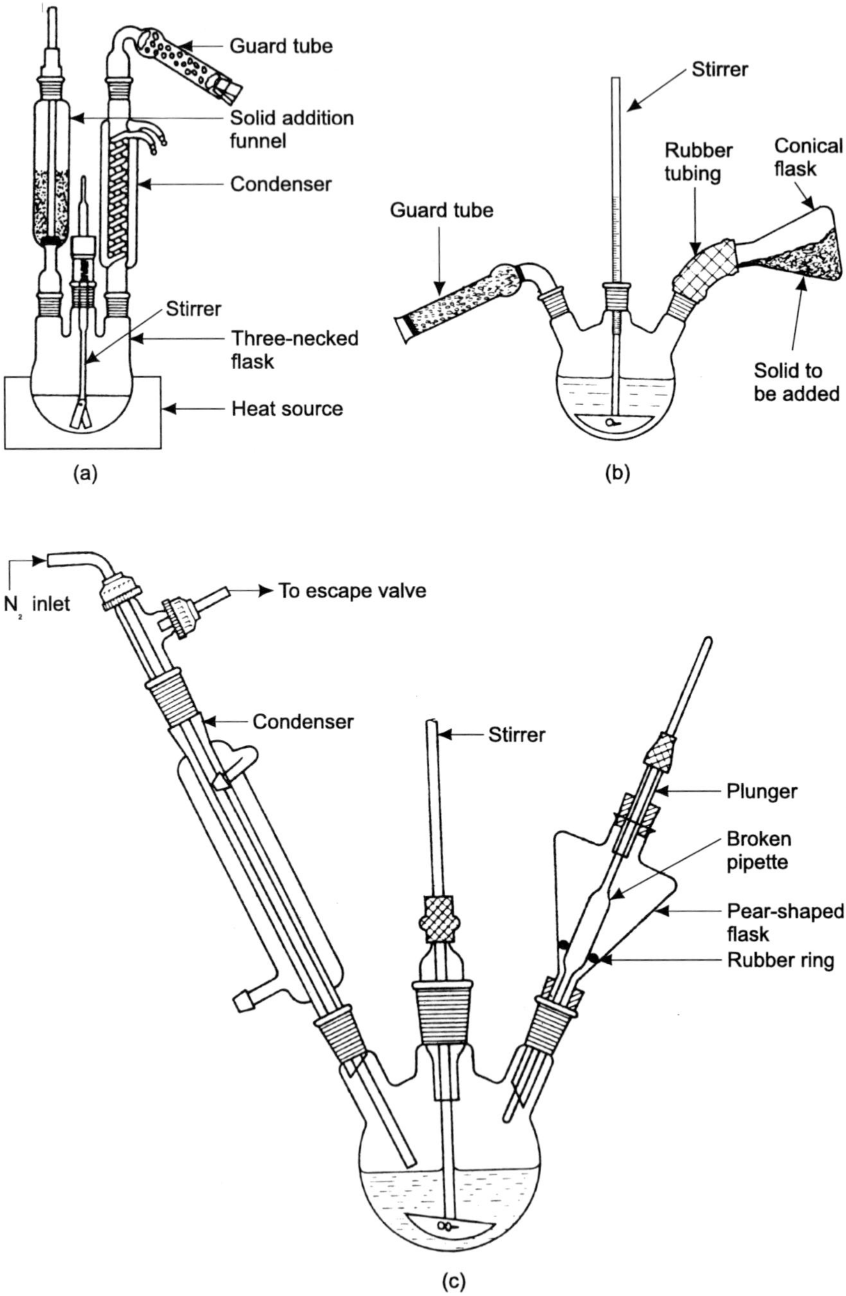

Fig. 1.30 Apparatus for a reaction under reflux with addition of a solid to a stirred solution in a dry atmosphere

cover the open end. If refluxing is not required, condenser can be omitted and the guard tube is directly fitted into the side socket of the three-necked flask.

In Fig. 1.30 (b), the solid is added in small portions by moving the flask up and the addition can be arrested by letting the flask to come down to its original position.

Fig. 1.30 (c) shows a conical reservoir made by a small pear-shaped flask and a broken pipette. Here also, the solid is added to the reaction mixture by lifting the plunger and the addition is stopped by moving it down so that the rubber ring covers the open end.

1.4.6 A Reaction Under Reflux with Addition of a Liquid and the Provision for the Absorption of any Gas Evolved in the Reaction

Fig. 1.31 (a-b) shows the assembly for the above reactions. Here, the liquid is added by dropping funnel, and (a) is the provision for the absorption of the evolved gas, e.g. if the evolved gas is water-soluble e.g. Hydrogen chloride or bromide etc. water is taken in the test tube. The receiver tube should be above the water (or any other liquid) in the test tube. The other way is to use inverted funnel as shown in Fig. 1.31 (b).

1.4.7 A Reaction Involving Addition of a Liquid with Distillation of a Volatile Component Simultaneously

Fig. 1.32 shows such type of reaction assembly.

In this assembly, the distillation unit i.e. thermometer, condenser, vacuum adapter and receiver flask is fitted to the main assembly of addition of liquid to reaction mixture with the help of a Claisen adapter.

This assembly can also be used for distillation of a large quantities of solvent from a liquid product, without changing the flask [as more the mixture can be added to the flask by dropping funnel].

1.4.8 A Reaction Involving Use of a Dean-Stark Trap

Fig. 1.33 illustrates the use of a Dean-Stark trap.This apparatus can be used for the azeotropic distillation with a solvent which is less dense than water, e.g. benzene, toluene.

Dean-Stark trap is used in those reactions where water is formed as a by product when such solvents are used. By using this trap, water can be removed from the reaction mixture simultaneously.

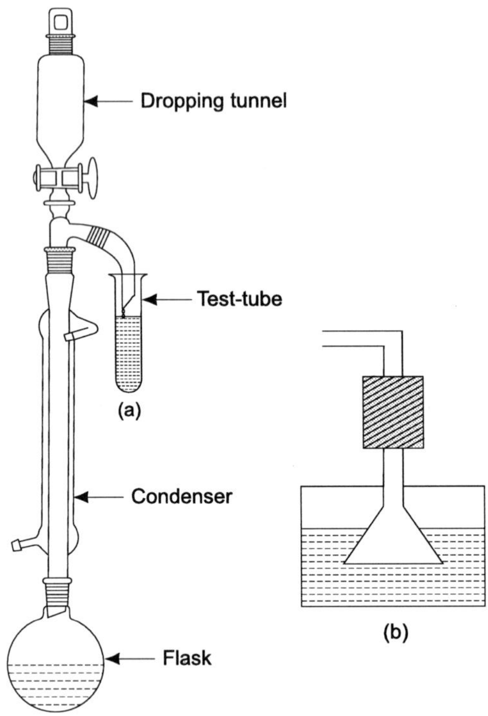

Fig. 1.31 Apparatus for a reaction under reflux with addition of a liquid and absorption of evoled gas

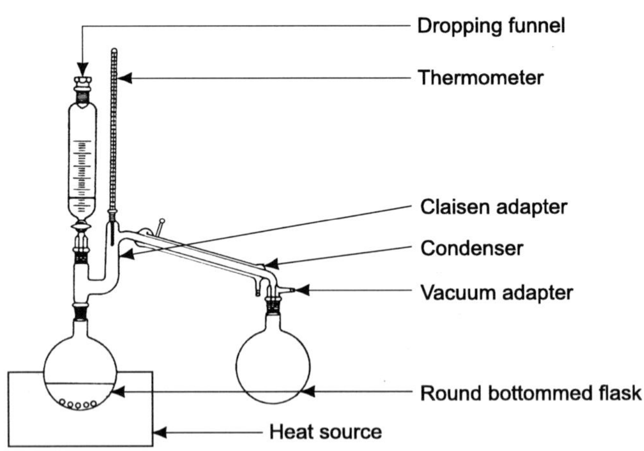

Fig. 1.32 Apparatus for a reaction under reflux with addition of a liquid along with the distillation of a volatile component

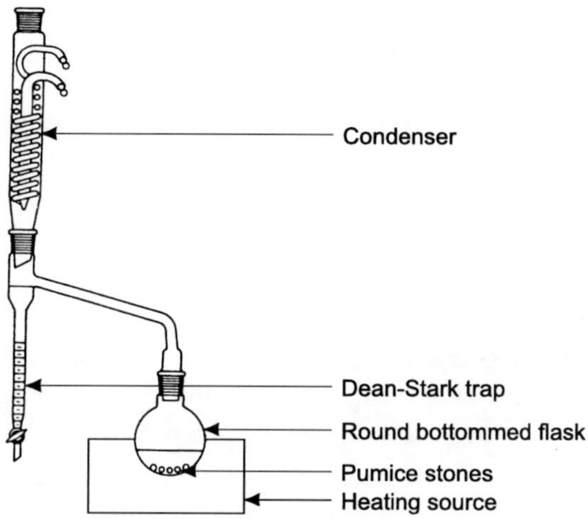

Fig. 1.33 Apparatus for a reaction using Dean-stark trap

1.4.9 A Reaction Under Reflux with Addition of a Gas to a Stirred Solution

Fig. 1.34 explains the assembly used for the above reaction.

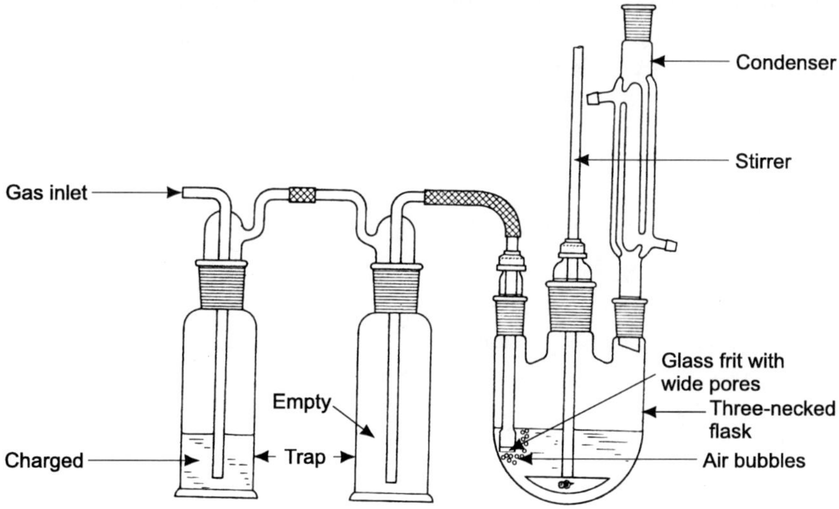

Fig. 1.34 Apparatus for a reaction under reflux with addition of a gas to a stirred solution

In Fig. 1.34, the gas is added under the stirred solution by using a glass tube fitted with a glass frit with wide pores. Advantage of using this glass frit is to improve the gas-liquid contact and help in the absorption of the gas. But it should not be used in those cases where the solid particles (either in the solution or as a product) may block its pores.

The gas should be passed into the solution via traps which are suitably charged to carry out the purification and drying of the gas before going to the reaction mixture and also traps prevent the damage to the cylinder (from which gas is available to the experiment) valve.

If the gases are generated chemically e.g. carbon dioxide, ozone gas should be maintained by having a suitable control on the generating apparatus. Proper attention should be given to such apparatus.

For the gases which are available in cylinders, the precautions for their use should be read carefully from the manufacturer's notes. Its antidote should be kept in such a place so that it can be used immediately when required. It should be noted down that the valve of the cylinder may be broken off when the cylinder is knocked and hence creating a highly dangerous situation. Therefore, the cylinders must always be supported in suitable stand carefully when in use or in storage. Cylinders of toxic gases, which are small in size, e.g. chlorine, sulphur dioxide etc. are to be kept in the fuming cupboard by an efficient support, while for cylinders of toxic gases, which are large in size, e.g. ammonia etc. are to be kept outside the lab on the laboratory wall so that gas can be passed to the reaction mixture in the fuming cupboard of the lab via the holes in the wall. It should be noted down that these cylinders (which are kept outside) should be placed in such a way that they get protection from weather etc.

1.4.10 A Reaction in Atmosphere of an Inert Gas

Certain reactions, e.g. reactions of organometallic compounds, are better carried out in an inert gas atmosphere to prevent the undesired side reactions due to the presence of oxygen and water vapour. The apparatus assembly for such type of reaction is shown in Fig. 1.35.

From the top of the condenser, dry nitrogen is inserted, which sweeps through the apparatus and then escapes out from the mouth of the dropping funnel, which has pressure equalising side tube. Rate of flow of inert gas is moderate initially and then reduced after a few minutes so that a slight pressure of gas is maintained (by adjusting the level of mercury in escape wall) throughout the reaction at a minimum flow rate.

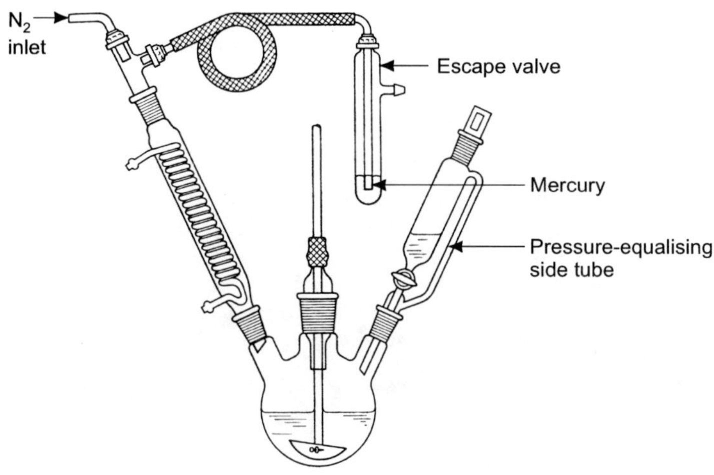

Fig. 1.35 Apparatus for a reaction in an inert gas atmosphere

1.5 STIRRING ARRANGEMENTS IN THE LABORATORY

In many organic preparations, the reaction mixture has to be stirred. A continuous stirring is required either during the addition of a reagent to mix it with the whole of the solution or throughout the preparation to increase the efficiency of heat transfer from a heating bath or to a cooling bath so that a constant temperature is maintained throughout the mixture.

For short time reactions, the manual stirring can be done and for the homogeneous solutions in which neither the exothermic nor the endothermic reactions are involved, i.e. not much heat transfer, then occasional stirring is sufficient. When in a reaction heat is generated locally which can decompose the product or a precipitate is formed by which adsorption is possible, continuous stirring is required. If continuous stirring is required for long, the solution is agitated by mechanical/ magnetic stirring. In many preparations, less time is required to complete the reaction, the temperature can be controlled easily and better yields are obtained.

There are two stirring techniques: one is mechanical stirring and another is magnetic stirring.

1.5.1 Mechanical Stirring

When stirring is required in beakers or conical flasks (open containers), a stirrer may be attached to a small electric motor directly. Fig. 1.36

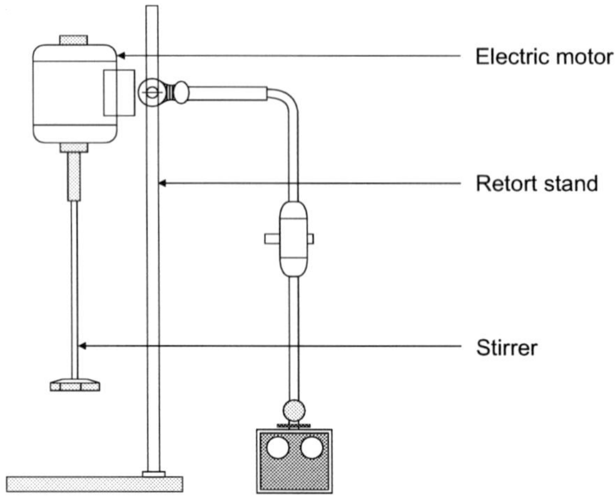

Fig. 1.36 Mechanical stirring in open container

illustrates such type of stirring. Glass stirrers are attached to motors by means of a small length pressure tubing (heavy-walled rubber tubing) while steel stirrers are attached with the screws. The speed of the motor and hence of the stirring is governed by a rheostat.

Usually glass stirrers are used in the laboratory. Stainless steel or Teflon stirrers may also be used. Teflon stirrers have the following advantages which make them important for use:

(1) They are soft and bend when hit the glass and so reduce chances of breakage of glass apparatus while using them.

(2) They can be shaped to fit the bottom of the flask and so make stirring of small volumes of liquid (or any container) in a large flask possible.

Fig. 1.37 (a-f) shows a variety of stirrers. Fig.1.37 (a-d) are various types of glass stirrers. An advantage of stirrer (d) is that it can be used for narrow necked containers also, while (e) is a Teflon stirrer with a half-moon shape. It has the advantages that it also can be used for narrow necked containers and stirring can be carried out in both round bottommed or flat bottommed container, by turning the blade over in the latter. (f) is a Herschberg stirrer which is used for vigorous stirring in a round bottommed containers and even of viscous mixtures. It is made up of a glass rod to which a glass ring is sealed which is then threaded by about 1 mm thick nichrome wire. This stirrer can also be used for narrow necked flask. One more advantage of this stirrer is that it can be used to stir the solid mixture which sticks to the bottom of the container.

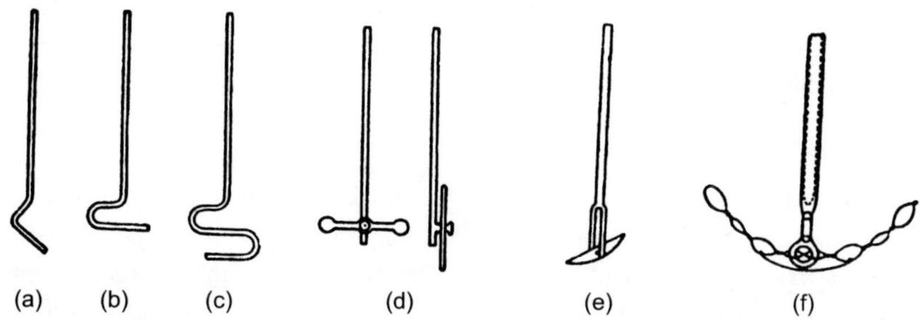

Fig. 1.37 (a-f) Various types of stirrers

Generally, the stirring of a solution, which is boiling under reflux is required and therefore the stirrer must be attached with a seal to prevent the escape of vapour. Fig. 1.38(a) shows a stirrer with a seal. Here, the stirrer is fitted quite closely within a short glass tube which passes through the cork of the flask. A short length of rubber tubing is fitted over the top of the tube to prevent the escape of the vapour. The stirrer is gripped strongly to the tube by turning back of over upper portion of the tube to the stirrer. And for the free rotation of the stirrer, few drops of mineral oil is applied between the stirrer and upper portion of the tube.

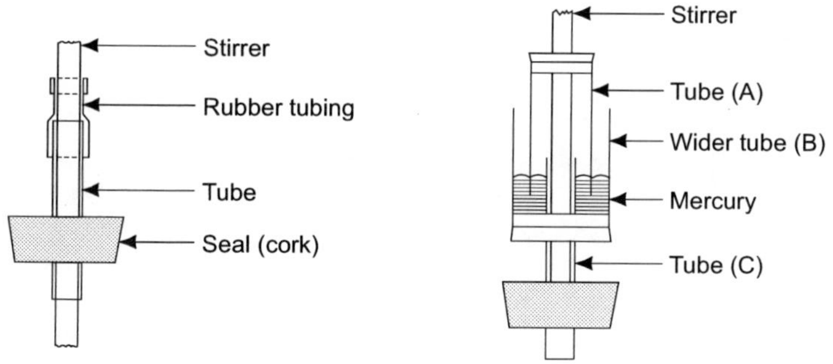

Fig. 1.38 (a-b) Stirrer with a seal

Another way is to use mercury sealed stirrer (Fig. 1.38 (b)). It consists of a short glass tube (C) which passes through the cork of the flask and the stirrer is fitted into it. A wider tube (B) is fixed to tube (C) with the help of a cork. While a similar tube (A) is fixed to the stirrer at the top. Tube (A) is well fitted in the space between tubes B & C which is filled with mercury to act as a seal.

1.5.2 Magnetic Stirring

Here, the stirring is carried out by using a glass or Teflon-covered bar magnet, stirring bar. The reaction flask having bar magnet is placed over a magnetic stirrer. Heating and stirring can be carried out simultaneously by means of a combination of magnetic stirrer hot plate. When the reaction flask is heated by the hot plate via oil bath, the stirring of reaction mixture can be done by the magnetic stirrer. Fig. 1.39 shows this type of arrangement.

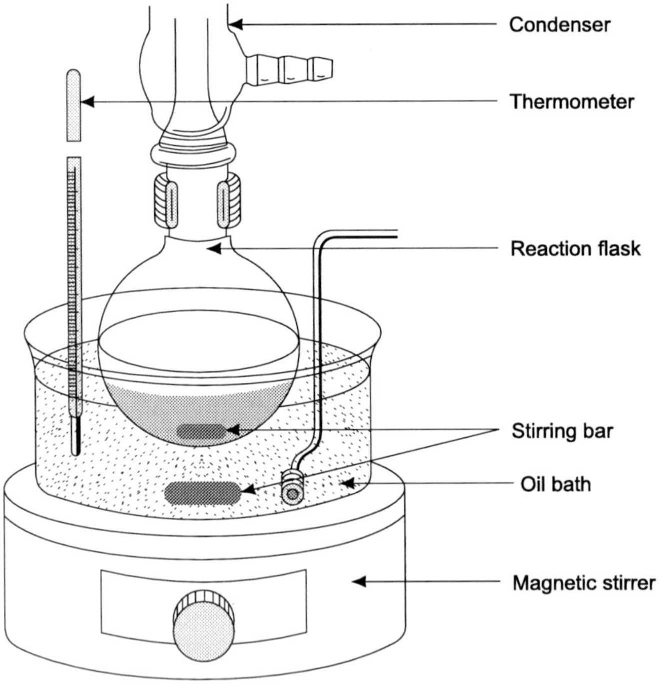

Fig. 1.39 Apparatus for simultaneous refluxing and stirring

The principle of magnetic stirring is shown in the Fig. 1.40. A bar magnet is attached to the shaft of an electric motor [magnetic stirrer] horizontally as shown in the figure. As the magnet of the motor turns, the magnet in the flask turns and the solution is stirred. the stirring speed is regulated with a rheostat. To start stirring, rheostat (in off position) is slowly rotated to attain the required rate of stirring. After completion of the reaction, rheostat is turned off to allow the stirrer to come to rest.

Magnetic stirring can be very conveniently used in closed systems e.g. when an inert atmosphere is being used, when an anhydrous environment is required, when small containers are used etc.

Mechanical stirring provides the greater torque require to stir the viscous mixtures, suspended solids or very large volumes. So in such cases, mechanical stirring is more advantageous than magnetic stirring.

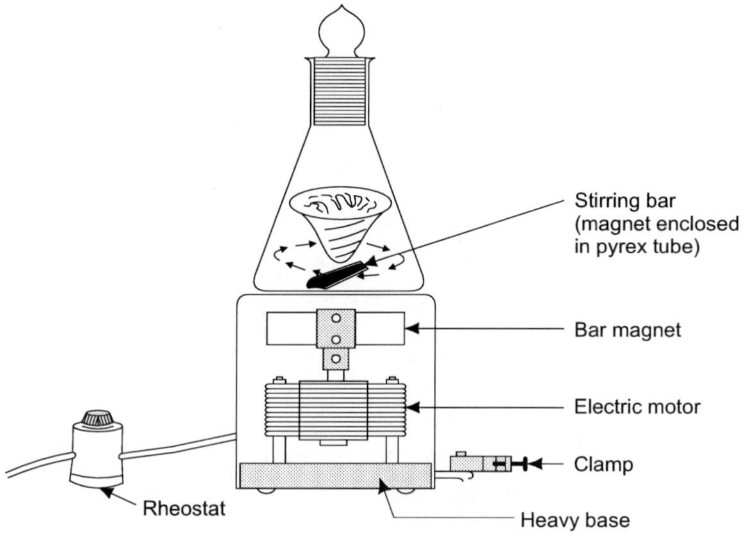

Fig. 1.40 Apparatus showing the principle involved in magnetic stirring

1.6 HEATING TECHNIQUES

Generally, the chemical reactions require heating or cooling. The following are the various techniques for heating the reaction mixture:

1.6.1 Heating with Bunsen Burner

In laboratory, the general method of heating is using a Bunsen burner flame. Rate of heating can be controlled by adjusting the height of the flame and the distance of the flask from the flame. The use of wire gauze in between flask and flame [as shown in Fig. 1.41] is to moderate and spread the heat of the flame. This way of heating has two advantages:

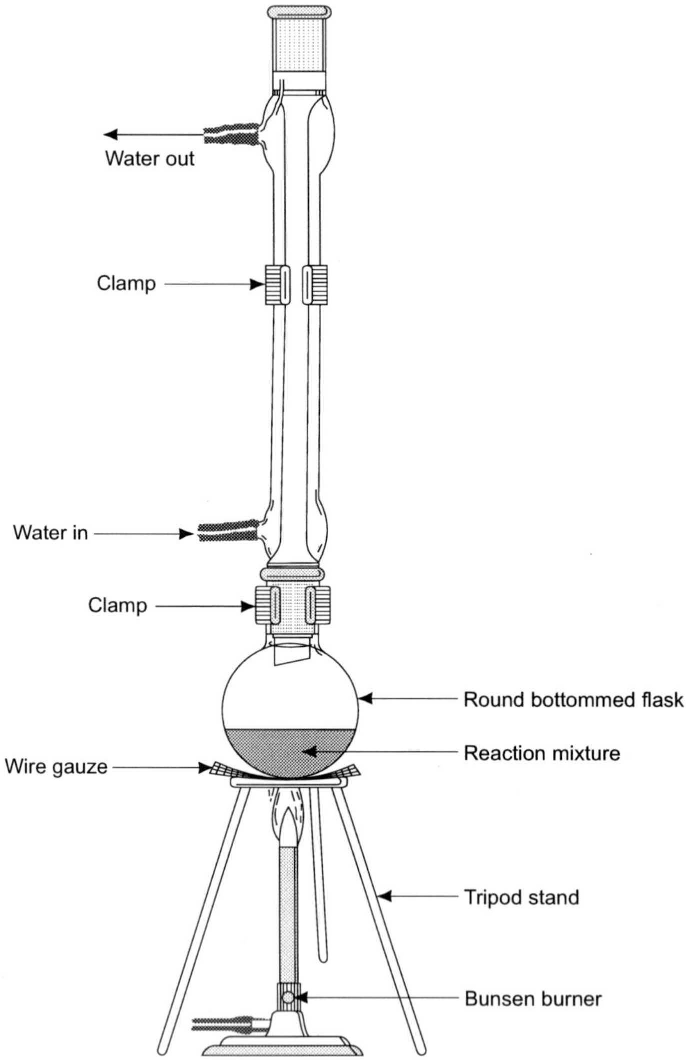

Fig. 1.41 Heating with a Bunsen burner

(1) Its set up is quick and easy.

(2) The heating of the mixture to the required temperature is very rapid.

While following are the disadvantages of this method:

(1) Heating is not uniform.

(2) The reaction mixture may set fire, if it boils over or the flask cracks or breaks during the reaction course, and

(3) The reaction mixture heated by the Bunsen burner flame require watching constantly as otherwise, the height of the flame may change to cause either too much heating so that reaction mixture boils over and catches fire or too less heating to boil the reaction mixture.

This method can be used satisfactorily, if small amount of reaction mixture is to be heated for a short period of time [constant watching is required].

1.6.2 Heating with Water Bath or Steam Bath

This source of heating is for temperatures up to 100°C. A water bath can be a beaker or any container having smooth coating of iron, containing water and is heated. If the water bath is heated with a flame, only non-inflammable liquids can be used, as the inflammable liquid may catch fire by this flame. For inflammable liquids, electrically heated water bath with a constant-level device may be used. Fig. 1.42 (a-b) shows the two types of water bath in which heating is carried out with an immersion heating element and heating is controlled by a regulator.

Fig. 1.42 (a) shows a circular type water bath which consists of a series of concentric rings to accommodate different sizes of flasks and beakers.

Fig.1.42(b) shows a rectangular type water bath with several holes having a series of concentric rings.

1.6.3 Heating with Sand Bath

A sand bath is a metal dish of sand, which is heated by a flame. It has the disadvantage that here the temperature cannot be controlled as the sand has low heat of conductivity, and so not used generally in the laboratory. This method of heating may be carried out in those reactions where high temperature is needed.

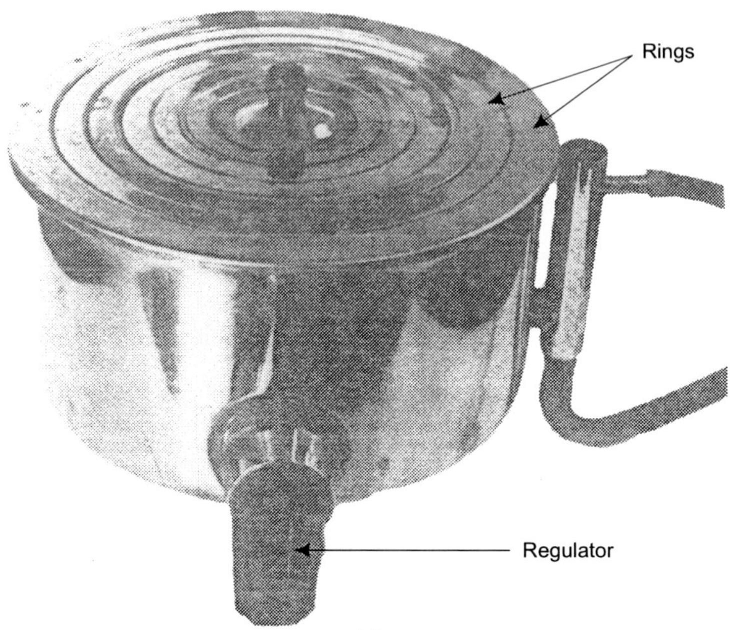

(a)
(circular type)

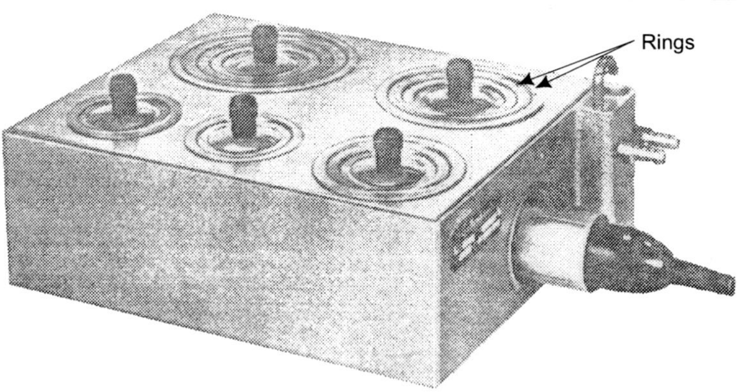

(b)
(Rectangular type with five holes)

Fig. 1.42 Types of water bath

1.6.4 Heating with Oil Bath

Oil baths are used for temperatures above 100°C. Various bath liquids used in oil baths are:

(a) Glycerol and dibutyl phthalate: It may be used for temperatures up to 140-150°C.

(b) Paraffin Oil: It may be used at temperatures up to about 220°C.

(c) Polyethylene Glycol: It can be used at temperatures up to 175-180°C. It is completely soluble in water and so cleaning is easy. While it is hygroscopic in nature and so on exposure to humidity for a long time, it starts boiling before the required temperature.

(d) Hard hydrogenated cotton seed oil: It has m.pt. 40-60°C and is non-sticky, clear liquid which solidifies on cooling.

(e) Silicone oils: They may be heated up to 250°C.

Oil baths should be kept in the fuming cupboard. To study the temperature, a thermometer should always be kept in the bath. After removing the flask from an oil bath, it is allowed to drain for few minutes and then the stuck oil is wiped with a cloth.

Disadvantages of the oil baths are :

(1) They start to smoke at high temperatures [e.g. above 180°C for polyethylene glycol, above 250°C for silicone oils].

(2) Temperature control often requires adjustment of heat source.

(3) If oil bath cracks or is broken down during the reaction course, oil may spills.

1.6.5 Heating with Metal Bath

For high temperatures, metal baths may be employed, e.g.

(a) *Wood's metal:* 4:2:1:1 parts of Bi:Pb:Cu:Sn [m.pt. 71°C]

(b) *Rose's metal:* 2:1:1 parts of Bi:Pb:Sn [m.pt. = 94°C].

(c) *Eutectic mixture of Sn* and *Pb:* 63 parts of Sn and 37 parts of Pb. [m.pt.=183°C].

They are solids at room temperature and do not smoke or catch fire.

Disadvantages of metal baths are:

(1) They are very expensive.

(2) Control of temperature needs often adjustment of heat source.

If heated above 350°C, rapid oxidation of metal alloy takes place, while using metal baths, precautions should be noted down that (a) The container should be either heated in a large luminous flame till carbon is deposited on whole of it or it is coated with graphite, to prevent the adhering of the metal on it, (b) the containers etc. dipped in molten metal must be removed from it before the metal solidifies.

1.6.6 Heating with Air Bath

This method of heating is used for non-inflammable liquids above 80°C. Air bath can be easily made by a cylindrical tin can, top of which is first smoothed and then a number of holes is made through bottom. A circular asbestos board of same diameter as that of the can and 2-3 mm thickness is then inserted over the holes. Asbestos cloth is then wrapped to the body of the can with the help of two wires near the top and bottom respectively. Another piece of a circular asbestos board of diameter slightly larger than that of the top of the can and 2-3 mm thick and having a hole in the centre to accommodate the neck of the flask is then cut diametrically as shown in the Fig. 1.43. An air bath placed on a tripod, is heated with a burner. The flask should not rest on the bottom of the bath and should be clamped as shown in Fig. 1.43.

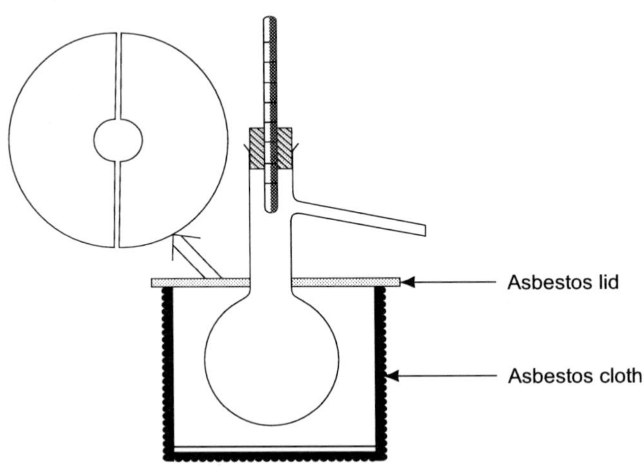

Fig. 1.43 Air bath

Various advantages of air bath are:

(1) Inexpensive apparatus, easy to construct.

(2) Control of temperature is easy.

(3) Inspection of flask may be carried out by removing asbestos clothes.

(4) The cooling of the reaction mixture may be done rapidly by completely removing the air bath.

1.6.7 Heating with Electric Hot Plates

It is also one of the methods of heating. The electric hot plate is having either a three heat switch [low, medium or high] or with a thermostatic control. The top of the hot plate is made up of heavy cast iron whose diameter varies from 5" to 10". Asbestos sheet should be placed between the metal top and the container to be heated to avoid bumping. They should not be used for the low boiling, inflammable liquids in an open container to prevent ignition caused by spilling of the vapours on to the heated metal.

An oil bath may be heated on a hot plate. This is an even and constant source of heating. The thermometer dipped in the bath is used to determine the temperature of the bath.

1.6.8 Heating with Heating Mantles

It is another method of electrical heating. Heating mantle is made up of aluminium metal in which a knitted glass-fibre fabric containing insulated nickel-chrome resistance wire is fitted. It is placed around the lower half of the flask of suitable size (which fits into it) as shown in the Fig. 1.44. The rate of heating is controlled by means of a variable voltage transformer. It is an easy and constant controlled source of heat and so the apparatus can be left unattended. Even risk of ignition is less as compared to the flame.

The various disadvantages of this method are :

(1) The flask-glass is at a temperature higher than the contents of the flask.

(2) The content level of flask should not fall below the top of the mantle as otherwise, the glass between the top of the mantle and that of the liquid will be at much higher temperature than the rest of the flask and so the material which splashes on the part will be superheated.

(3) Heating with heating mantle is a slow process and takes much time to reach thermal equilibrium and the cooling of the contents is a slow process as even after the turn off the heat, the contents of the flask are boiled as mantle has high heat capacity value. And therefore, for rapid cooling, the mantle has to be removed.

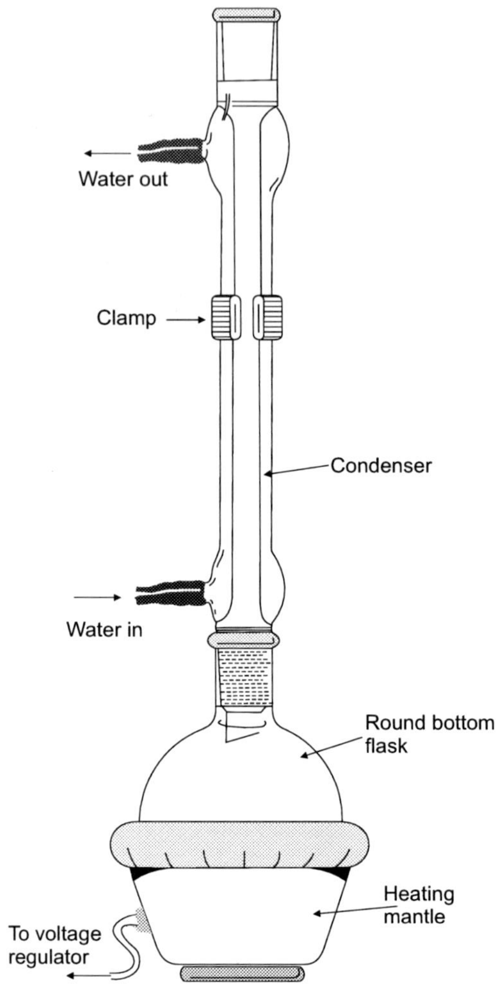

Fig. 1.44 Heating with a heating mantle

(4) For different sizes of flasks, different heating mantles are required because the flask must fix into the mantle.

1.7 LOW TEMPERATURE EXPERIMENTS

A suitable cooling bath is used depending upon the temperature of the flask to be cooled. Various cooling baths are listed below:

1.7.1 Ice-Water bath

A mixture of ice and water is used for 0-5°C temperature. Crushed ice gives better results.

(a) For cooling to a few degrees below 0°C, freezing mixture can be used which is made by thorough mixing of 1 part of sodium chloride and 3 parts of crushed ice. This freezing mixture lowers down the temperature from – 5° to – 10°C.

(b) Another ice-salt mixture is made by 5 parts of $CaCl_2.6H_2O$ and 3.5-4 parts of crushed ice to give cooling from –40° to – 50°C.

Disadvantages of these ice-salt baths are that they are short-lived and inefficient.

1.7.2 Dry Ice/Solvent bath

Dry ice (solid carbon dioxide) may be used for very low temperatures. For efficient cooling for long periods, these are the best cooling baths. Mixtures of dry ice with various solvents are:

(a) Dry ice with chloroform or carbontetrachloride lowers down the temperature to – 50°C.

(b) Dry ice with ethylalcohol to – 72°C.

(c) Dry ice with acetone and diethyl ether to – 77°C.

These baths are prepared by adding lumps of Dry Ice to the solvent until the required temperature is obtained. At any point of time, the temperature can be maintained by adding Dry Ice. It should not be added too rapidly as it will result in the overflow of bath on the evolution of carbon dioxide.

Mercury freezes at – 40°C and so an alcohol or toluene thermometer is to be used at very low temperatures.

When the cooling is required for long periods, a freezing mixture is to be kept in a glass or copper thermos flask.

1.7.3 Refrigerator

It may also be used as a cooling chamber.

1.8 PREPARATION OF GASES

1.8.1 Carbon Dioxide

Carbon dioxide (> 99% purity) is available in cylinders of compressed CO_2 and can be procured from the market. Besides solid CO_2 (dry ice) is also available and can be used for the generation of CO_2 gas.

However, in the laboratory pure CO_2 can be prepared in the usual Kipp's apparatus (Fig. 1.45). Marble chips are taken in the middle compartment and dil. hydrochloric acid (1:3) is added from the top. The carbon dioxide generated is washed by passing through a wash bottle containing water.

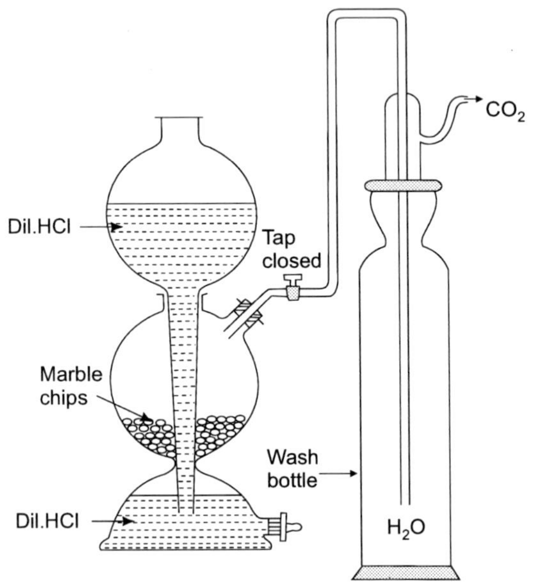

Fig. 1.45 Kipp's apparatus for the preparation of CO_2

The carbon dioxide gas thus prepared can be used for Kolbe's reaction and for precipitation of phenolic compounds from their solutions in sodium hydroxide solution.

1.8.2 Hydrogen Gas

Hydrogen (99.9998% purity) is available in cylinders of compressed hydrogen and can be procured from the market. In this the impurities comprise traces of nitrogen, oxygen, moisture and hydrocarbons. For catalytic reductions, the hydrogen is quite suitable and no purification is usually necessary.

In the laboratory, hydrogen gas is prepared in the Kipp's apparatus (Fig. 1.46)

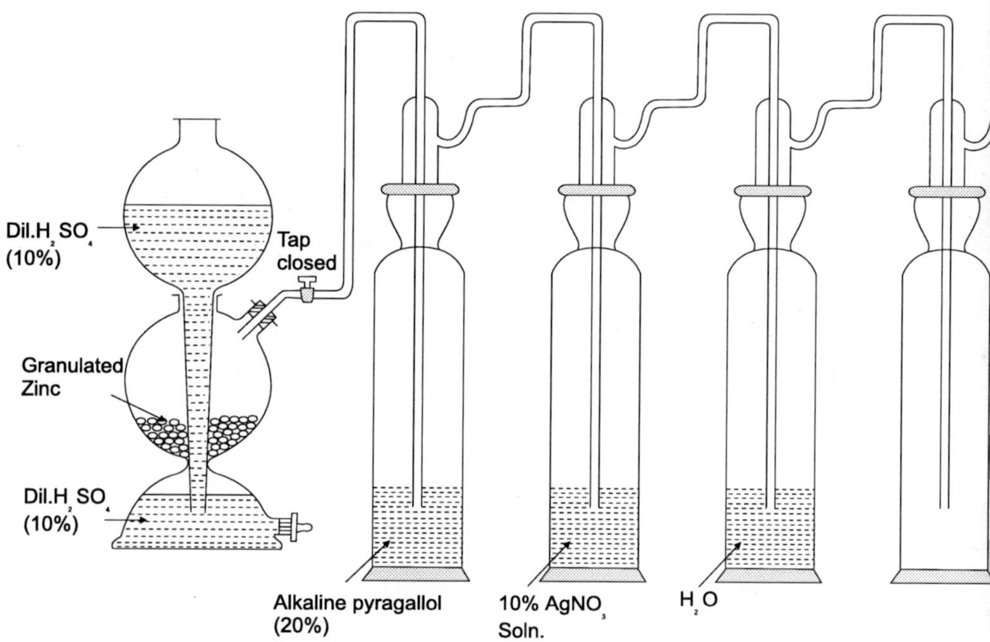

Fig. 1.46 Preparation of Hydrogen

The hydrogen produced is passed successively through wash bottles containing alkaline pyrogallol solution (20%, 100 ml NaOH solution containing 20 g pyrogallol), silver nitrate solution (10%), water and finally through empty wash bottle. This hydrogen thus produced is suitable for most of the reactions.

1.8.3 Hydrogen Bromide

It is most conveniently prepared by the action of bromine on tetralin (tetrahydronaphthelene) (Fig. 1.47).

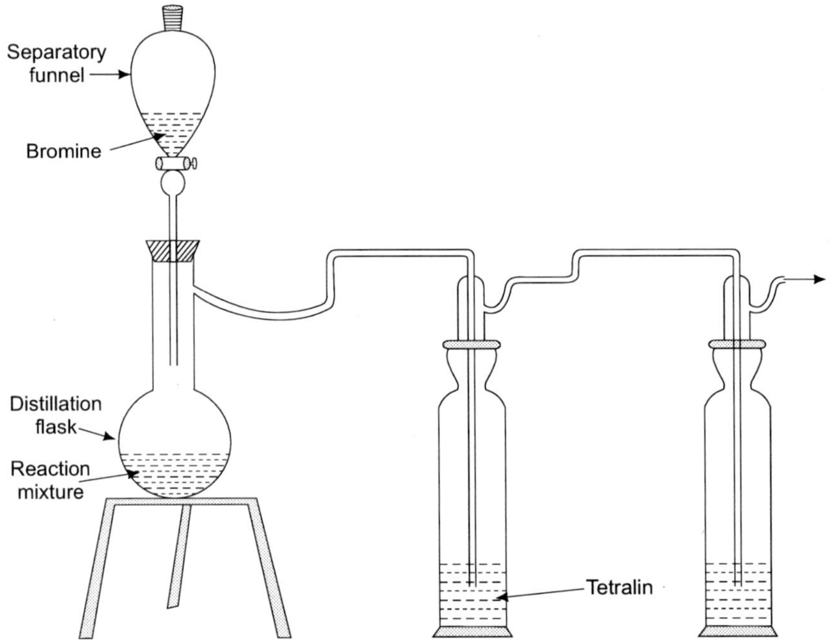

Fig. 1.47 Generation of Hydrogen bromide

As seen, only half the bromine is converted with hydrogen bromide. Dry tetralin (Sec 5.1.48.) is taken in a distillation flask which is filled with a separatory funnel. Bromine is allowed to drop in from the funnel at a regular rate and the contents of the flask are generally swirled from time to time to ensure a steady evolution of hydrogen bromide. The generated hydrogen bromide is passed through a wash bottle containing tetralin (which removes traces of bromine carried over by HBr). Finally the gas is passed through an empty wash bottle to prevent back flow. The evolved HBr may be used either as such or passed into glacial acetic acid (the solution of HBr in acetic acid is used).

1.8.4 Hydrogen Chloride Gas

Method (i) From commercial sodium chloride and conc. sulphuric acid

Sodium chloride (commercial) is taken in a 2 litre filtration flask. It is just moistened with conc. hydrochloric acid. Concentrated sulphuric acid is added drop-wise through a separatory funnel fixed with the help of a rubber stopper at the mouth of the filtration flask. The hydrogen chloride gas obtained is passed through a wash bottle containing conc. sulphuric acid and finally through an empty wash bottle. This gives pure dry hydrogen chloride gas in continuous supply (Fig. 1.48).

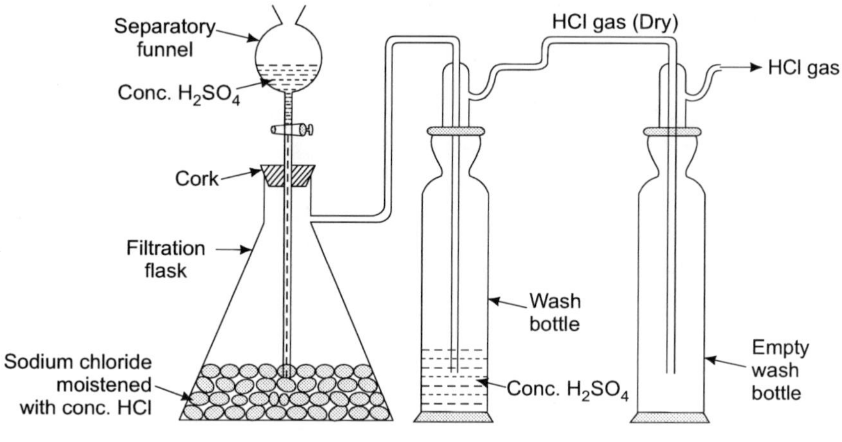

Fig. 1.48 Preparation of Dry Hydrogen Chloride Gas

The dry hydrogen chloride gas prepared in this way is suitable for Hoesh reaction, Gatterman reaction and other related reactions.

Method (ii) From concentrated hydrochloric acid and concentrated sulphuric acid

The apparatus used is the same as used for the preparation conc. sulphuric acid and commercial sodium chloride. About 150 ml conc. sulphuric acid is placed in the filtration flask and 100 ml conc. hydrochloric acid is taken in a separatory funnel and the apparatus as shown in the figure 1.49. The stem and the separatory funnel dips into conc. H_2SO_4. Conc. hydrochloric acid is added carefully through the separatory funnel. The liberated HCl gas is dried by passing through a wash bottle containing conc. H_2SO_4 and finally through an empty wash bottle to prevent the vapour of H_2SO_4 going into the reaction and also for prevention of back flow. The yield is 31-32 g of hydrogen chloride per 100 ml of concentrated hydrochloric acid.

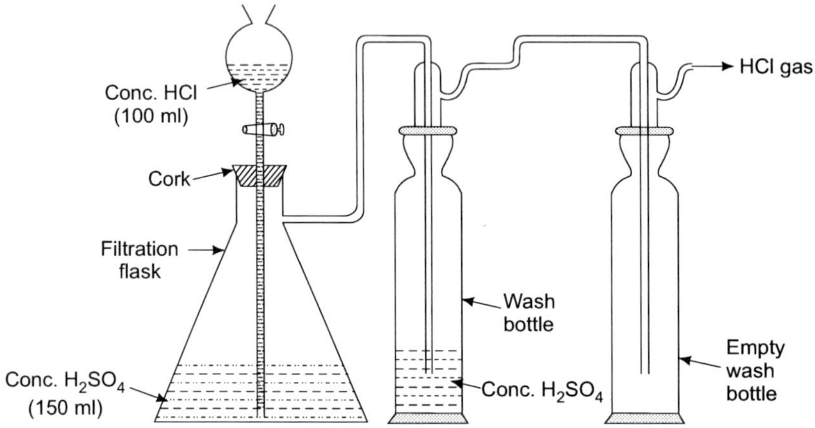

Fig. 1.49 Preparation of hydrogen chloride gas

1.8.5 Hydrogen Sulphide Gas

It is prepared in the Kipp's apparatus (Fig. 1.50).

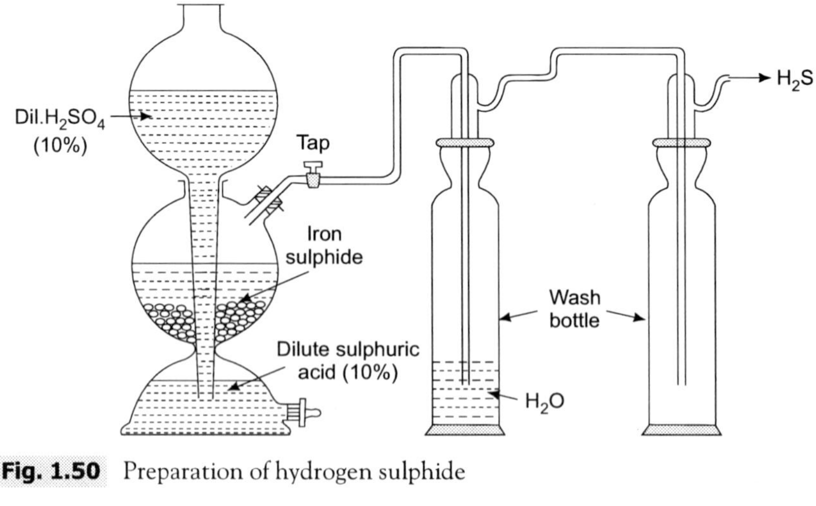

Fig. 1.50 Preparation of hydrogen sulphide

Ferrous sulphide (small lumps) is taken in the middle compartment and diluted sulphuric acid (10%) is added from the top. The hydrogen sulphide produced is passed through a wash bottle containing water and finally through an empty wash bottle to prevent acid vapour passing along the gas and also to prevent back flow.

Caution: It is poisonous in nature.

1.8.6 Sulphur Dioxide Gas

Method (i) A continuous supply of sulphur dioxide can be obtained by taking copper turnings in a distillation flask fitted with a separatory funnel (Fig. 1.51). Conc. sulphuric acid is added slowly through the separatory funnel while the distillation flask is heated on a wire gauze. The sulphur dioxide generated is washed by passing through a wash bottle containing water and finally through an empty wash bottle.

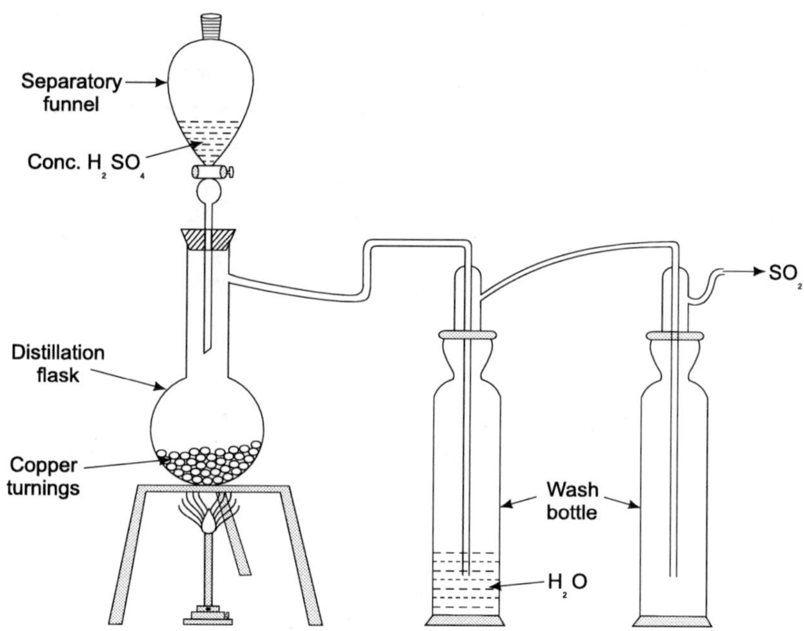

Fig. 1.51 Preparation of sulphur dioxide

Method (ii) Sulphur dioxide is best prepared by taking sodium sulphite and water in the form of a paste in a distillation flask (See Fig. in method (i)). Conc. hydrochloric acid is added through a separatory funnel while the distillation flask is heated on a wire gauze. The sulphur dioxide gas is bubbled through a wash bottle containing water and finally through an empty wash bottle. This procedure is used for the preparation of small amount of sulphur dioxide.

1.9 DRYING OF ORGANIC COMPOUNDS

Before performing any reaction, the reactants (and the solvent if used) must be dried. Even the presence of very small amount of water in a reaction mixture is undesirable. However, if the reaction has to be performed in aqueous medium, drying of reactants and the solvents is unnecessary. Following are given the methods for drying (removal of water) from solids, liquids and gases.

1.9.1 Drying of Solids

1.9.1.1 Use of air-ovens

The sample is heated in an air oven at a temperature reasonably below the melting point of the solid. For high melting solids, a temperature of $< 110°C$ is used.

A small amount of a compound can be dried by using the arrangement as shown below (Fig. 1.52). In this the sample to be dried is taken in a small sample tube which is kept in a slanting position in a comparatively bigger tube (e.g. a boiling tube). A small amount of glass wool is placed at the bottom of the tube and the tube is connected to vacuum. It takes about 4-6 hrs. for removing the moisture.

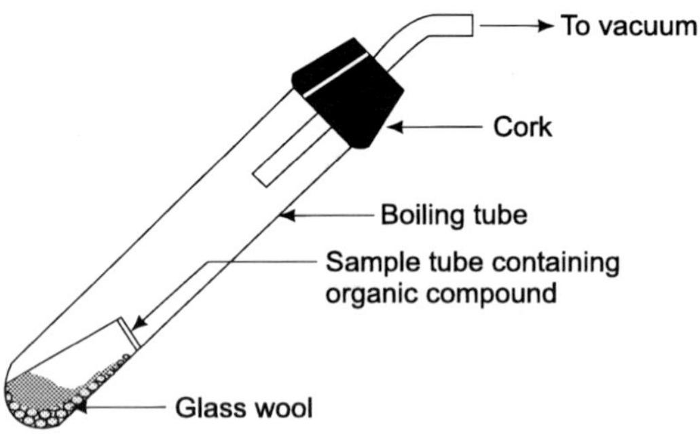

Fig. 1.52 Drying of small sample

1.9.1.2 Use of vacuum desiccator

The organic compound can be made free of moisture by drying in a vacuum desiccator (see Fig. 1.53). The organic compound is put in a petri dish on the selt of the desiccator. A suitable desiccant is put in the lower compartment and vacuum is applied through the stopper. It takes about 6-8 hrs for drying.

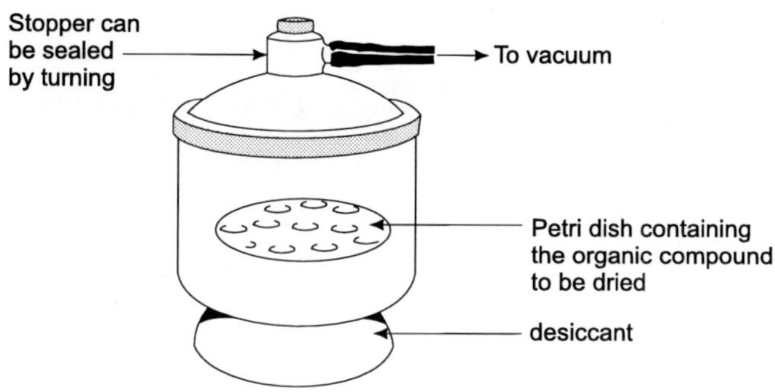

Fig. 1.53 A Vacuum Desiccator

Depending on the nature of the compound to be dried, the desiccant can be aluminium oxide, Conc. H_2SO_4, KOH pellets, anhydrous $CaCl_2$ or drierite ($CaSO_4.\frac{1}{2} H_2O$). [See Section 1.3.14.1].

1.9.1.3 Use of a drying pestle

In this process the organic compound is dried at comparatively higher temperature in vacuum. The apparatus used is shown below (Fig. 1.54). The drying pestle consist of a double surface tube having a B24 (Male Joint) at the bottom and B24 (Female Joint) at the top. The organic compound to be dried is kept in a tube which is slided into the pestle. At the bottom, a R.B flask (250 ml capacity) is placed which is heated on a heating mantle.

The flask contains a solvent whose boiling point is at least 10-15° below the melting point of the solid to be dried. On the top, a reflux condenser is fitted. Vacuum is applied and heating continued. It takes 3-4 hrs for complete drying. The procedure is used for drying compounds for micro analysis.

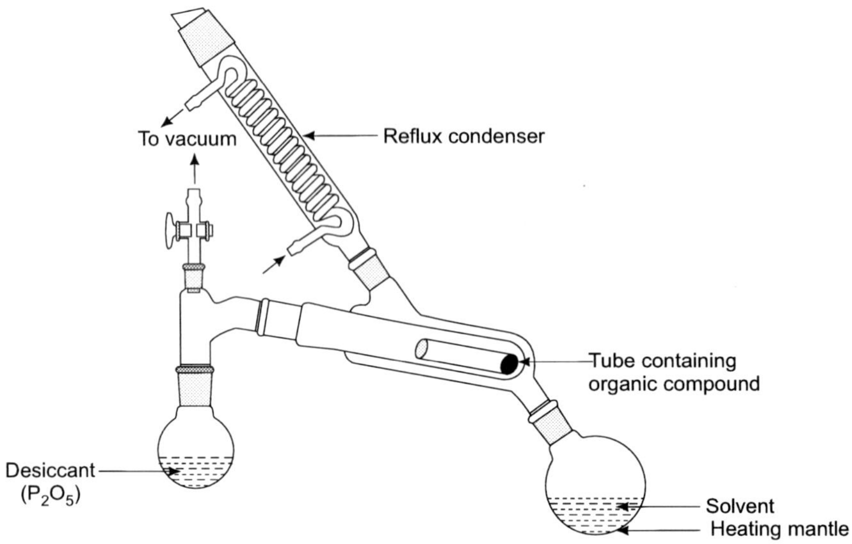

Fig. 1.54 A Drying pistol

1.9.1.4 *Use of infrared heat lamp*

This method is useful for drying compounds (especially products obtained in organic reactions). In this method the powdered compound is spread on

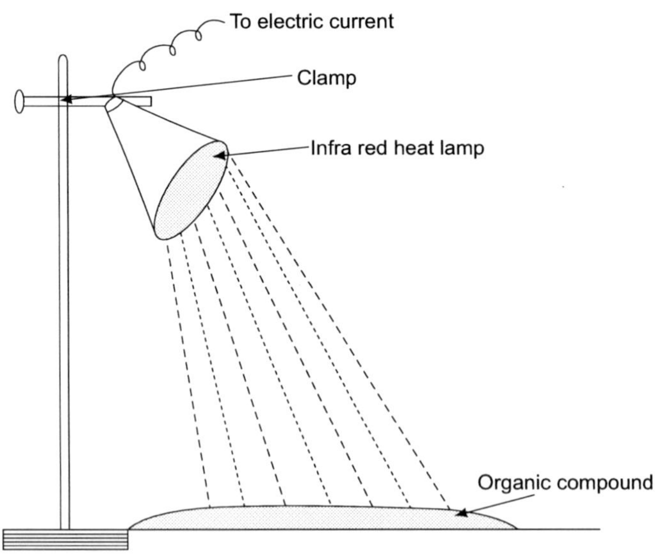

Fig. 1.55 Infra red heating lamp

a sheet of filter paper and the I.R. heat lamp is put on the top at a distance of about 12-16" fixed on a stand (see Fig. 1.55).

1.10 REMOVAL OF SOLVENTS

The removal of a solvent from a solution of organic compound can be carried out by distillation. Sometimes an organic compound may be decomposed at these high temperatures of distillation and whole of the preparation get spoiled.

Rotatory film evaporator (shown in Fig. 1.56) can be employed for the removal of solvent. Here, evaporation of the solvent is carried out under reduced pressure using a water pump, and so low temperature for evaporation. The evaporator flask is heated on a water bath along with rotation for thorough mixing and prevention of bumping. The solvent rises to the spiral condenser, gets condensed and collected into the receiving flask. More of the solution is added into the evaporator flask through feed tube.

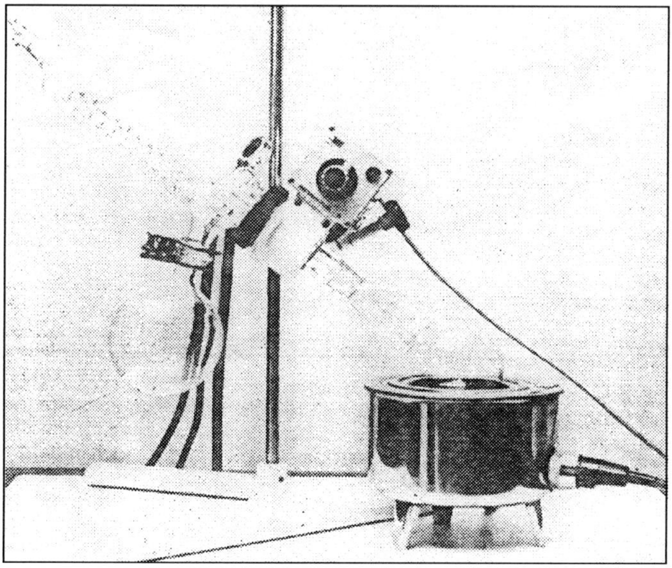

Fig. 1.56 Rotatory film evaporator

2

Purification of Organic Compounds

A pure substance contains only one kind of molecules while an impure substance is a mixture of different kind of molecules. Thus to purify an impure substance is to separate the desired molecules from the mixture and therefore, the purification is a separation. The differences in the properties of different molecules are the basis of various separations. Products of organic reactions are seldom pure; they may be contaminated with either the starting material or side-products. To obtain the satisfactory physical constants, the substance must be pure. And to carry out the purification, various techniques have been employed depending on the physical state of the compound, which are discussed in this chapter.

2.1 CRYSTALLIZATION

Crystallization is a simple, effective and very important technique to separate and purify solids. It is based on the fact that all organic compounds are more soluble in hot than in cold solvents, so that solid gets dissolved on heating and is obtained back on cooling.

2.1.1 Steps Involved in Crystallization

The following steps are involved in the purification by crystallization:

(a) Selection of a solvent
(b) Dissolution of the sample
(c) Decolourisation of the solution
(d) Hot filtration
(e) Cooling for crystallization
(f) Cold filtration
(g) Washing the crystals
(h) Drying the crystals
(i) Checking the purity

(a) Selection of solvent

The selection of solvent is very important for purification by recrystallization. The solvent for crystallization should have the following characteristics:

(i) Solid to be purified must be insoluble in the cold and highly soluble in the hot solvent.

(ii) Impurities must be either readily soluble or insoluble in solvent.

(iii) Solvent should yield well-formed crystals of the purified compound.

(iv) It should have low boiling point so that it can be easily removed from the purified compound.

(v) It should not react chemically with the compound.

The following generalisations, with large number of exceptions, may help in the choice of a solvent for recrystallization:

(1) Like dissolves like i.e. a polar substance is more soluble in polar solvents and less soluble in a non-polar solvents. The order of increasing polarity of solvent is: cyclohexane, light petroleum, carbon tetrachloride, benzene, diethylether, chloroform, acetic acid, ethyl acetate, acetone, alcohol, methanol and water.

(2) A solvent will most readily dissolve compounds, which are similar in chemical and physical properties with it, e.g. alcohol will usually dissolve other hydroxy compounds, benzene will dissolve hydrocarbons etc.

(3) In ascending a homologous series, the solubilities become more and more like the hydrocarbon from which they may be derived.

If two or more solvents are found to be equally suitable for recrystallization, the final selection should depend on the inflammability and cost of the solvent. The common solvents employed for the recrystallization are shown in Table 2.1.

The choice of solvent should not be made on the basis of theoretical consideration alone but should be determined experimentally, if no information is available. About 0.1 g of the powdered compound is placed in a clear dry test-tube, 1 ml of the solvent is added drop-wise with continuous shaking, if the compound is readily soluble in cold, it is unsuitable. If it does not dissolve, warm the mixture gently by taking precautions of inflammability of the solvent. Addition of solvent is

Table 2.1 Common solvents for Crystallisation

S.No	Solvent	b.p. (°C)	Remarks
1.	Diethylether*	35	Inflammable
2.	Petroleum ether	Boiling fractions 40-60 60-80 80-100 100-120	Inflammable
3.	Dichloromethane**	41	Non-inflammable & toxic
4.	Acetone	56	Inflammable
5.	Chloroform**	61	Non-inflammable & toxic
6.	Methanol**	64.5	Non-inflammable & toxic
7.	Carbon tetrachloride**	77	Non-inflammable & toxic
8.	Ethanol	78	Inflammable
9.	Ethyl acetate	78	Inflammable
10.	Benzene***	80	Inflammable, highly Toxic and Carcinogenic
11.	Cyclohexane	81	Inflammable
12.	Acetic acid	118	Not very inflammable and pungent vapours

* Its use should be avoided wherever possible because of its high inflammability and also its tendency to creep up the walls of the containing vessel and hence results in the deposition of solid material by complete evaporation rather than the crystallization.

** Its vapours are toxic and therefore inhalation should be avoided and the recrystallization should be done in fuming cupboard.

*** Toluene is much less toxic than benzene and so wherever possible it is used preferentially.

Use of carbon disulphide, b.p. 46°C, should be avoided as it has a very low flash point and its vapours form explosive mixture with air.

continued in 0.5 ml portion each time, if the substance is not dissolved completely. The mixture is heated after each addition. If even after adding 3 ml of solvent, the compound does not dissolve on heating, the solvent is unsuitable due to the low solubility of the compound in it. If an almost clear solution is obtained, the tube is cooled by immersing it in cold water and if the solution is not clear, it is filtered and then the tube is allowed to cool. If crystallization does not start rapidly, it may be due to the lack of nuclei for crystal growth and therefore the scratching of the tube below the surface of the solution is done with the help of a glass rod. Crystals often form rapidly after scratching as fine scratches on the walls form the sites for crystal growth. If crystals do not separate even after scratching another solvent should be tried. If the crystals separate out, its yield should be noted. The same procedure is repeated with various other possible solvents using a clear fresh test-tube each time, to select the best solvent. Note down carefully the approximate proportions of the solute and solvent for having the best results of crystallization.

If the substance is highly soluble in one solvent and highly insoluble in another solvent, crystallization can be carried out by mixed solvents, provided the two solvents must be completely miscible, e.g. ether/ n-hexane, benzene/petroleum ether, alcohol/water, acetic acid/water etc. First, the compound is dissolved in the solvent in which it is highly soluble and another solvent is added drop-wise till turbidity appears. The former solvent is added drop-wise to the mixture till the turbidity has just disappeared and then it is heated and filtered and allowed to cool, crystals will come out to form a clear.

(b) Dissolution of the sample

After finding a suitable solvent and the approximate proportions of solute and solvent for crystallization, a solution of the sample in the hot solvent can be prepared.

If the solvent is neither inflammable nor toxic or inexpensive, the recrystallization can be carried out in a conical flask and if the solvent has any of the above characteristics then the recrystallization is done in round-bottommed flask with a condenser to avoid danger and wastage of the solvent. Figure 1.25 shows the typical apparatus for recrystallization.

The size of the flask should be chosen such that it will be less than half filled with the solution. Place the solid in the flask and add slightly less than the required amount of solvent. Pumice stones must be added in the flask to prevent bumping, a reflex condenser is then fitted on the flask and the mixture is heated to boil. For solvents having boiling point below 80°C, water bath or steam bath is used and for higher boiling solvents, hot plate or sand bath is used. More solvent is added to the mixture through the condenser, till a clear solution is obtained. If near the end of the dissolution process, an additional amount of the solvent does not dissolve any more of the solid, then this remaining solid is likely to be an insoluble impurity. If the solution then obtained is not clear, it is filtered.

(c) Decolourisation of the solution

The presence of coloured impurities in the sample or in a solution of a colourless compound is quite obvious. During crystallization process, these coloured impurities get dissolved in the boiling solvent and adsorbed by the crystals, giving an impure product. Such type of impurities are preferentially adsorbed by the activated charcoal so can be removed by boiling the coloured solution with a small amount of activated charcoal for 5-10 minutes followed by hot filtration. Now the filtrate is usually almost colourless and therefore pure crystals are obtained.

Not only the coloured impurities can be removed by the activated charcoal but also the resinous or finally divided matter can also be removed from the solution. Sometimes the solution is not clear due to the presence of resinous or very finely divided insoluble impurities, which cannot be removed by simple filtration. These impurities can also be removed by heating the solution with activated charcoal. The activated charcoal takes away these from the solution and makes it clear.

Only a small amount of activated charcoal must be used (1-2% by weight of the sample) otherwise some of the compound which is to be purified may also get adsorbed. The same procedure could be repeated with some more fresh charcoal (1-2%), if this amount is not sufficient enough. It should also be noted that the activated charcoal should be added only after the solution is cooled otherwise the solution would boil over when it is added. After this addition, the solution can be reheated to the boiling point.

Animal charcoal contains large proportion of calcium salts and should not be used with acidic solutions and also the impurities present in charcoal itself may get dissolved in hot solvent and separate out with the crystals of the compound. These problems can be sorted out by boiling animal charcoal under reflux with dilute hydrochloric acid (1:1) for 3 hours. Then the mixture is diluted with hot distilled water and filtered through a Buchner funnel. It is washed repeatedly with hot distilled water till all the acid has been removed, drained and finally dried by heating in an evaporating basin in an electric oven.

(d) Hot filtration

Now the desired compound is in the solution in the hot solvent along with insoluble impurities and decolourising carbon, if used. They can be separated by gravity filtration. Vacuum filtration cannot be used as the reduced pressure in the suction flask will boil the filtrate and the solid will get deposited on the walls of the flask.

In hot filtration, main problem is that undue cooling takes place, which results in the crystallization of some of the solid during the filtration. This can be reduced to a minimum by using heated funnel with no stem. A stemless funnel avoids the problem of crystallization and hence clogging in the stem. Therefore to avoid the undesired crystallization, the filtration is carried out through a fluted filter paper* supported in a heated, stemless funnel and pouring only a little of solution at a time and keeping the remainder at the boiling point. If the solid separates out on filter paper during filtration, it must be scraped back into the flask, redissolved and then filtered.

If very large amount of hot solutions are to be filtered, the funnel should be warmed externally during the filtration (hot water funnels). The hot water funnels are shown in Figure 1.8.

The filtrate should be reheated to dissolve any crystals, which are formed during filtration, to give a clear solution.

(e) Cooling for crystallization

After making the filtrate clear by dissolving any crystals formed during

---

*Preparation of a fluted filter paper.

The filter paper is first divided into eight equal sectors by making fourfolds in the paper. The folding is continued, the edge 2,1 is then folded on to 2,6 and 2,3 to 2,5 to produce new folds 2,7 and 2,8 respectively. Further 2,3 to 2,6 and 2,1 to 2,5 give new folds 2,9 and 2,10 respectively. Finally a fold is made on each of the eight segments e.g. between 2,3 and 2,9/2,9 and 2,6 and so on in a direction opposite to the first series of folds to give a fan arrangement which on opening gives a fluted filter paper. The above sequence is shown in the Figure 2.1

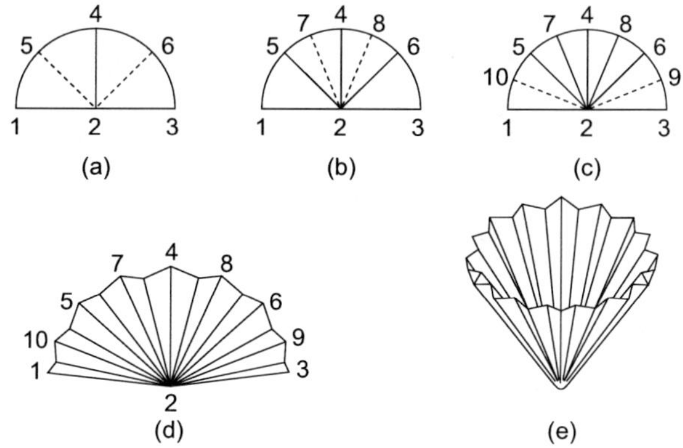

Fig. 2.1 Preparation of a fluted filter paper

crystallization, the filtrate is covered with a watch glass and kept aside to cool undisturbed.

Rate of cooling determines the size of crystals. Slower cooling gives large crystals while faster cooling gives smaller crystals. Very large crystals are undesirable as they *often occlude* the solvent and its dissolved impurities. Very small crystals are also not favourable as it is difficult to wash them free of solvent and soluble impurities and take much time to dry. The best quality of crystals is usually obtained by allowing the solution to cool to room temperature on a non-heat conducting surface. If small crystals are required, the hot solution is stirred in a cold water or ice bath. The rate of cooling can be slowed down by keeping the hot filtrate flask in the beaker containing water at the temperature of it and then both are allowed to cool to room temperature.

(f) Cold filtration

When crystallization is complete, the product is collected by suction filtration using a Buchner funnel or Hirsch funnel (later is used in those cases where only a small quantity of solid material has to be filtered). The size of the funnel used should be such that not more than half of it will be filled with crystals while the suction flask should not be filled with the solution above the tip of the funnel or side arm.

The Buchner & Hirsch funnel is already explained in Fig.1.9(a-b). The Buchner funnel is fitted by means of a rubber cone into the neck of a thick-walled filtering flask, suction flask (Fig. 1.10) which is then connected to a similar flask, trap, by means of rubber pressure tubing, which is then connected to filter pump by pressure tubing. Use of a trap is very essential as water can be sucked back and added up in filtrate if there is sudden fall in water pressure. The filtration through suction is rapid and satisfactory method to remove the solution from the crystals. A well-fitted Whatmann paper (No. 1) [it should cover the entire perforated plate and should have the diameter slightly less than the inside diameter of the funnel] is placed in the Buchner funnel and is wetted with a few drops of the liquid to be filtered, so that it can attached firmly to the funnel when suction is started and avoids the loss of solid material from passing down to the flask. The mother-liquor is decanted through the funnel until the mass of crystals left in the flask is just covered with the solvent. The entire mass of crystals is then poured into the funnel by swirling the flask so that crystals get suspended in the solvent and then they are poured quickly into the funnel.

These processes (swirling and pouring) are repeated till whole of the solid material has been transferred into the funnel. The suction is continued till most of the liquid is filtered.

The above procedure (filtering the bulk of mother liquor followed by the entire compound) is quicker than bringing the solid material into suspension and then filter it. And also, a gentle suction is more effective and rapid filtration than powerful suction as the finer particles of the solid may drag into the pores of the filter paper in the latter case and hence results in the slower filtration.

(g) Washing the crystals

Crystals are now washed with some fresh cold solvent to remove soluble impurities from the surface of the crystals, otherwise crystals will be contaminated with these impurities when solvent get evaporated.

If the crystals are relatively soluble in the solvent, then a minimum amount of cold solvent must be used. While if they are not very soluble in solvent, washing can be done by large amount of solvent (no need to cool it). If the crystals are not in the form of solid cake, washing can be done by releasing the vacuum and then pouring of the wash liquid over the crystals and again applying the vacuum, while if the solid cake is formed, the solvent is added to the crystals in the funnel and the solid is carefully broken down and suspended by the help of a spatula. Care should be taken during the suspension that the entire solid product gets suspended without the tear or dislodge of the filter paper. Alternatively, the solid cake can be transferred to a breaker and then washing is done by breaking the cake in the solvent and the product is again collected by suction filtration.

When a non-volatile solvent e.g. acetic acid is used for recrystallization, the washing of the crystals is done with a more volatile solvent, in which crystals are not soluble to wash off non-volatile solvent so as to speed up the drying process.

(h) Drying the crystals

The crystals obtained by suction filtration can simply be dried by spreading it on the thick pad of filter paper with the help of a spatula. Another similar pad is placed on the top and then the solid is pressed. These sheet pads are changed by the fresh pads occasionally as the mother liquor wets them.

Further drying can be carried out in the electric oven provided the compound is stable. The temperature of oven must be maintained at 50°C to avoid the decomposition of the compound. For the removal of common organic solvents, 1-2 hr is sufficient at this temperature while water can be removed from the sample when it is kept in an oven overnight at this temperature.

The main disadvantage of drying through filter paper is that the recrystallized compound may be contaminated with filter paper fibre.

Another method to increase the rate of drying the sample is to use heat lamp. The infrared heat lamp is shined on the sample. The thinner the layer in which sample is spread, faster it will dry. The rate of recondensation of water can be decreased by circulating air over the sample.

The final drying is preferable carried out in a desiccator, if time permits. With the low melting solids, this is the best method of drying the sample. The drying agents employed for water absorption are granular calcium chloride, conc. sulphuric acid or silica gel. Silica gel having blue colour is used when it changes to red—it is regenerated by heating. Granular calcium chloride may also be used for absorption of methanol and ethanol vapours. While for absorption of vapours of ether, chloroform, carbon tetrachloride, benzene, toluene etc. paraffin wax is used as a drying agent.

For more efficient and rapid drying, vacuum desiccator (Figure 1.22, Ch-1) is used. When it is used, the substance is taken in a watch-glass covered with another inverted watch-glass to avoid the sweep away of finer crystals on sudden blow of air in the desiccator. One more precaution must be taken into account in this method of drying and that is a trap must be used in between the desiccator and pump, during exhausting a desiccator, to avoid the rush back of a mixture of air and water into the desiccator due to the cease working of pump for any reason.

For drying larger quantities, vacuum ovens are used i.e. sample is drying under vacuum at high temperatures.

(i) Checking the purity

To obtain the pure compound, it may require repeated recrystallization. The purity is checked by melting point determination and until a sharp melting point is obtained, the recrystallization process must be repeated.

2.1.2 More Techniques of Crystallization

Till now, the normal techniques for recrystallization have been discussed. However, various problems may come up during recrystallization e.g. separation of liquid or oil product resulting in the failure to get crystals etc. Such problems encountered during the recrystallization are discussed in this section.

(a) Separation of liquid product

Sometimes the product in crystallization separates as a liquid or oil instead of crystals. Oil is contaminated with impurities and so if solidifies on standing, sometimes considerable period is required for crystallization, impurities separate out along with crystals. So as soon as oil starts separating out, its solidification is carried out by either of the methods:

(i) Reheat the mixture to make it a clear solution and allow to cool it spontaneously and stir the mixture vigorously till oil is well dispersed in the solution and compound crystallizes out. Stirring may be stopped when all the oil has disappeared.

(ii) Addition of a small quantity of the crude product to seed the solution may cause the crystallization of the oil.

(b) Failure to crystallize

Sometimes due to the presence of various substances or tar, substances form supersaturated solutions and cannot crystallize out as these impurities act as a protective colloid. In such cases, following methods may be applied to facilitate the crystallization:

(i) By scratching the flask with a glass rod. This causes roughening of the surface, which induces the crystallization.

(ii) By sending the solution. The seed crystals are obtained by cooling the small quantity of solution in a suitable freezing mixture. These crystals initiates the crystallization and it continues till equilibrium is reached.

(iii) By cooling the solution in a freezing mixture. As cooling is done at very low temperature and so in some cases, crystallization can be hindered by the separation of viscous solid. In such cases, the mixture is warmed very slowly to attain optimum temperature required for the formation of crystals. Once even minute crystals separate out, at a slightly higher temperature, the size of crystals will increase.

(iv) By addition of solid carbon dioxide. On addition of a few lumps of it, cold spots are produced in the solution which facilitate the crystallization.

(v) In some cases, when all the above methods fail, the crystallization is carried out by keeping the solution in a refrigerator or freezer for long periods of time.

2.1.3 Semimicro Microrec Rystallizations

Sometimes very small quantity of a compound may be needed to recrystallize e.g. the compounds obtained by natural products, small-scale expensive preparations or by-products of reactions.

(a) Semimicro recrystallization

In this recrystallization the apparatuses used are shown in Figure 1.10 (b-c). The compound to be recrystallized is dissolved in semi-micro test tubes or centrifuge tubes. Heating should preferably be done in a water or oil bath to regulate the heat. While for inflammable solvents, solution is prepared under reflux conditions by using semimicro glass apparatus. The solutions are allowed to cool and separated crystals are filtered using a small Hirsch funnel as shown in Figure 1.10 (b).

Filtration nail may also be used. It is a thin glass rod flattened at one end, which is then fitted into a small glass funnel as shown in Figure 1.10 (c). The nail head is covered with a circle of filter paper of appropriate size. The size of the funnel and filtration nail may be varied depending upon the amount of crystals to be filtered.

(b) Micro-scale recrystallization

When micro quantity of the compound is to be recrystallized, usual semi-micro techniques of recrystallization give great loss of the crystals.

An easy and cheap way to recrystallize the compound on micro scale is as follows:

A test tube filled with only about 1/8th by the solution should be chosen and the sample is added to it. Solvent is added and the test tube is warmed in a water bath or on a sand bath. As the amount of compound is very small, heating should be done for a brief period only. Relatively large size of test tube helps in condensation of solvent vapours by providing large surface area.

The solution is now filtered into another test tube, smaller than the previous one, by using a Pasteur pipette with a small piece of cotton just at the point where the pipette starts to narrow. Keep it on the tip of the second test tube as shown in Figure 2.2. Second Pasteur pipette is used to transfer the hot solution from the test tube to the filter pipette. The solution passes down through the pipette to the second test tube, which can be warmed carefully in water bath or sand bath to keep the solution warm. After filtration, the second test tube is kept aside for cooling.

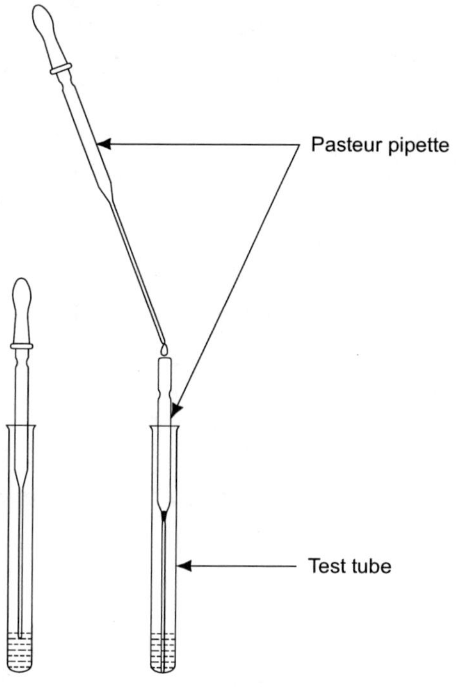

Fig. 2.2 Filtration in a micro-scale recrystallization

When crystallization is completed, the solvent can be removed by using another Pasteur pipette. It withdraws solvent by pushing its tip in the bottom of the test tube. Washing can be done by using cold solvent.

The solvent can also be removed by centrifugation. After centrifugation, crystals are settled down and the solvent can easily be decanted. Washed with the solvent and removed it by centrifugation followed by decantation.

The crystals are dried by connecting it to the vacuum as shown in the Figure 2.3.

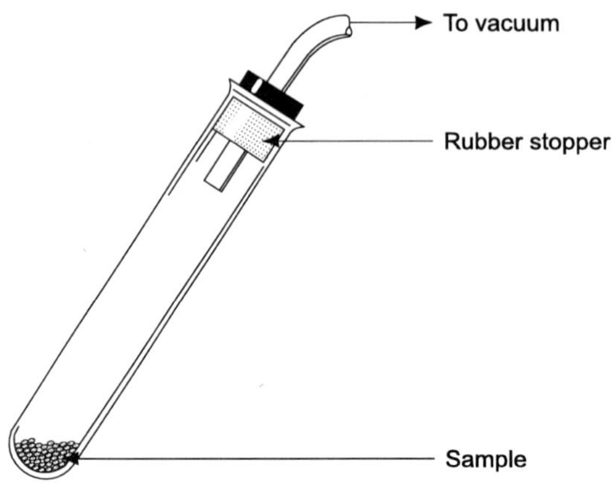

Fig. 2.3 Drying crystals in a test tube

Dried crystals are collected over a piece of filter paper by inverting the test tube and tapping it with a glass rod.

2.1.4 Exercises in Recrystallization

A few examples are given to understand the recrystallization technique.

(a) Benzoic Acid from distilled water (a non-inflammable solvent)

Some other compounds which can be recrystallized from water are acetanilide, salicylic acid etc.

Weigh 1 g of crude Benzoic acid into a 100 ml conical flask. Add about 20 ml of hot distilled water. Heat the mixture on sand bath or electric hot plate. Add more of hot water in small portions with continuous stirring and heating until the entire solid has dissolved. If the solution is colourless, filter the hot solution through a fluted filter paper supported on a stemless funnel. During filtration, keep the remaining solution hot with the help of a hot plate or filter the solution by using a hot water funnel (Figure 1.8). If the solution is coloured, allow it to cool slightly and then add about 0.1 g of decolourising carbon and boil the mixture for few minutes to remove the coloured impurities and then carry out the filtration as described earlier.

Collect the filtrate in a 100 ml conical flask, heat it again to dissolve unwanted crystals of the benzoic acid and then cover it through watch glass and allow to cool to room temperature of its own. After an hour when complete crystals separate out, filter them with suction using a small Buchner funnel. Wash the crystals twice with 5 ml portions of cold water to remove the mother liquor adhering to it. Press the crystals in the funnel with spatula to remove the mother liquor as much as possible. Remove the funnel from the filtration set and invert it on a pad of 3-4 filter papers resting upon another pad of newspaper and let them dry in air. For rapid drying, the crystals may be kept in oven on a watch glass. Note the yield and m.pt. of the recrystallized sample. The crystallization may be repeated if the recrystallized sample is not pure i.e. if compound melts over a range. Melting point of pure benzoic acid is 121°C.

(b) Naphthalene from alcohol (an inflammable solvent)

Some other compounds which can be recrystallized with an inflammable solvent are:

(a) m-dinitrobenzene from alcohol

(b) tribromoaniline from alcohol

Weigh 1 g of commercial naphthalene into a 50 ml round bottommed flask. Add 5 ml of rectified spirit and 2-3 pieces of pumice stone in it to avoid bumping. Fit a reflux condenser (Figure 1.25). Heat the mixture in a water bath or electric hot plate to boil the solvent. Add 2-3 ml of the solvent, boil carefully after each addition until whole naphthalene apart from insoluble impurities, has dissolved. If it is coloured, use 0.1 g of animal charcoal as described in the above procedure. Filter the hot solution through a fluted filter paper. Care should be taken that no flames must be there in the vicinity. Collect the filtrate in conical flask, heat it and cover it by a watch glass. When the complete crystals are separate out, filter them through a small Buchner funnel. Wash the crystals twice with 5 ml portions of cold rectified spirit. Press the crystals through spatula and dry them as described in the above procedure. Record the yield and melting point of the recrystallized compound. The melting point of pure naphthalene is 80°C.

(c) Sulphanilic acid from water

Crude sulphanilic acid is usually almost black in colour. It is recrystallized from distilled water and decolourized by using animal charcoal.

Weigh 1 g of crude sulphanilic acid in a conical flask and add 20 ml of distilled water. Heat to dissolve almost whole of the compound by using sand bath or hot plate. Cool it slightly and add 0.1 g of animal charcoal to the solution and continue boiling for few minutes. If the filtered solution is coloured, boil it with a further 0.1 g of animal charcoal. Filter it as in the above procedures and collect crystals of sulphanilic acid. Record* the yield.

(d) Para-aminoazobenzene from solvent pair

Aminoazobenzene (C_6H_5–N=N–C_6H_4–NH_2) is highly soluble in methylated spirit and insoluble in water. So for recrystallization this solvent pair is used. Dissolve the crude sample in boiling methylated spirit using water bath. Remove the conical flask from water bath and then add water drop by drop until the solution becomes just cloudy. Put the solution on a water bath until it becomes clear and then remove it from water bath immediately. Filter if necessary; allow it to cool to room temperature slowly. Filter the crystals and note the yield and melting point. The melting point of pure para-aminoazobenzene is 126°C.

2.2 FRACTIONAL CRYSTALLIZATION

Fractional crystallization is employed to separate a mixture of two or more substances in their pure states, provided they differ in solubilities in a particular solvent. The solvent to be used can be found out by trial and if there is no appreciable difference in their solubilities, the process of separation is difficult.

In this method, the mixture is treated with a minimum quantity of the hot suitable solvent to dissolve almost all of the more soluble part leaving the less soluble behind. The solution also contains a small amount of less soluble part. Now, the solution is filtered while hot and the more soluble substance present in the solution, is obtained on cooling. The solid, hence obtained, consists of the major portion of more soluble substance and a very small portion of less soluble substance. So, the solid is dissolved once again in the minimum amount of the solvent and recovered by crystallization. This whole procedure of making solution and crystallization is repeated a number of times until the pure component (more soluble) is obtained. Purity of the substance is checked by melting point determination.

* Melting point cannot be recorded as it decomposes on heating and therefore, it has no melting point.

The residue obtained after the first filtration consists of mainly the less soluble part along with a very small amount of the more soluble part. This (less soluble part) can also be obtained in the pure form by working up in the similar manner, the purity of which is also checked by melting point determination.

2.2.1 Exercise on Fractional Crystallization

(a) Separation of benzoic and cinnamic acid

The two acids differ in their solubilities in warm water (40°C). Cinnamic acid is insoluble in water at this temperature and benzoic acid is soluble. And, therefore, these two acids can be separated by fractional crystallization using warm water (40°C) as a solvent. A mixture of 2 g each of the above acids is placed in a 100 ml beaker and 50 ml of warm water (40°C) is added to it. The mixture is stirred for 5 minutes on a hot water bath. The temperature of the mixture is to be maintained at 40°C. The contents of the mixture are allowed to settle down and the supernatant liquid is filtered while hot using a hot water funnel. The filtrate is collected in another beaker. The residue left in the first beaker, is treated similarly with warm water six times and then the final residue is recrystallized from boiling water. The product cinnamic acid hence obtained, is checked for purity by melting point determination (i.e. 133°C). All the filtrates are combined in 400 ml beaker and is concentrated down to 50 ml. On cooling, benzoic acid is crystallized out, which is further purified by recrystallization from hot water. The purity of benzoic acid is also checked by melting point determination (i.e., 121°C).

2.3 DISTILLATION

If a sample of an organic liquid contains impurities, it may be purified by distillation. This method may also be used to determine the boiling point (4.3.1). If impurities are non-volatile, on simple distillation, they are left behind in the distillation flask and when the impurities are volatile, fractional distillation method is used to purify the given liquid.

In this section, simple distillation is discussed. The method is used when the pure liquid and impurities have widely different boiling points (b. pt.) (The b.pt. difference between the two must be at least 50°C). Typical apparatus used for simple distillation is shown in Figure 2.4.

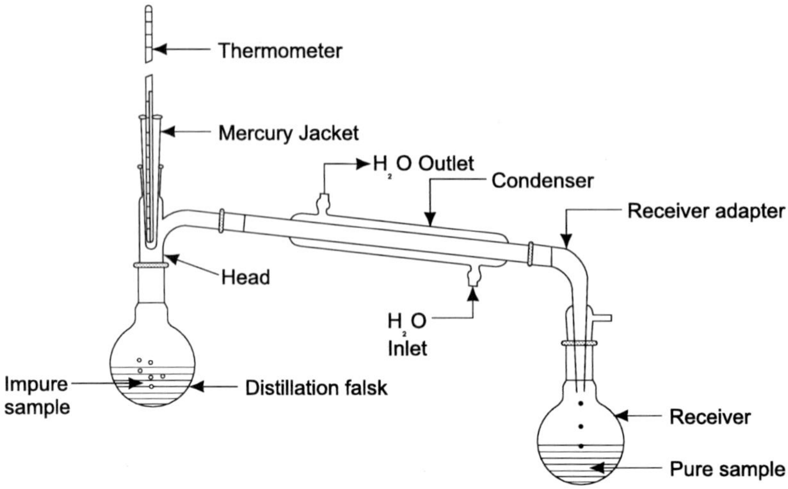

Fig. 2.4 Apparatus used for simple distillation

A distillation flask of suitable size is fitted to a water condenser, to which an adapter is fitted so as to collect the distilled liquid into a receiver. Thermometer is fitted in the neck of the distillation flask by means of a well-bored cork. The bulb of the thermometer should be in centre and just below the level of the side tube otherwise the accurate boiling point will not be obtained.

The following points must be taken into account, while carrying out the distillation process:

(1) The size of the distillation flask should be such that it will be one half or at the most two-thirds filled by the liquid. If the flask used is too large, then superheating and sometimes, decomposition may occur.

(2) Few pieces of unglazed porous porcelain or carborundum must be added in the distillation flask to provide a surface on which bubbles of the vapours can be formed and hence results in the gentle boiling. In the absence of these boiling stones, the liquid often reaches a temperature above its b.pt. and leads to the violent eruptions—'bumping'.

(3) The complete distillation apparatus should be tightly fitted as it may be used for inflammable liquids.

(4) All the clamps used should be lined with cork as otherwise glass apparatus may have cracks on applying excessive pressure on the clamp.

(5) In water condenser, the inlet of water should be from the lower end and outlet of water should be above the jacket so that the condenser is full of water.

(6) A water condenser is replaced by an air condenser [a glass tube with no jacket or emptying the water from condenser jacket] if the boiling point of liquid is above 150°C.

(7) For low boiling inflammable liquids, heating should be done by a water bath while for liquids having b.pt. more than 80°C, sand bath or direct heating on a wire-gauze can be used. Electrically heated mantles can also be used for all the liquids.

The liquid is poured in the distillation flask with the help of a funnel having stem which extends below the side arm, pumice stones* are added to it and a thermometer is placed in the neck of the flask. The flask is heated by a suitable means, described above. The temperature will first rise rapidly until it is near boiling point of the liquid, then it will rise slowly and then will become constant. This constant temperature is the boiling point of the liquid. So heating can be fast until boiling commences, then the flame must be decreased so that the one or two drops of the liquid are collected quickly. The distillation should not be conducted too slowly as otherwise due to the lack of a constant supply of vapour on the thermometer bulb will result in an irregular boiling point while it should not be too fast so that it causes superheating by heating directly a part of the vapour as well as liquid. If the liquid is pure, most of it will be collected at a constant temperature, i.e. boiling point of the liquid, while if two components are to be separated, the first component will pass over at a constant temperature, then before second component starts coming, an intermediate stage will come in which there is rise in the temperature, it is collected in another receiver and finally the second component will come over again at a constant temperature which is collected in the third receiver. It must be noted that never overheat the flask to distil the last drop of the remaining liquid – a small residue must be left in the flask.

For small quantity liquids, pear-shaped flask may be used. The apparatus is shown in Figure 2.5. A 10 or 25 ml pear-shaped flask having a

* Should never be added to the hot liquid.

long side arm is fitted with a condenser, which in turn is fitted with a receiver. The flask is heated in an air bath and distillate is collected in an appropriate size receiver.

If instead of attaining a constant temperature, there is a steady rise in temperature, simple distillation cannot be used to purify the sample. And for the purification, fractional distillation must be used.

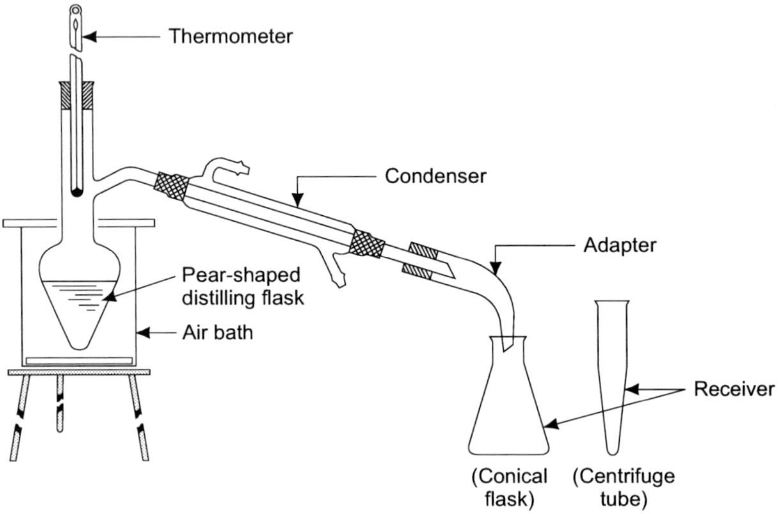

Fig. 2.5 Apparatus for micro-scale distillation

2.4 FRACTIONAL DISTILLATION

This method is used to separate the liquids whose boiling points difference is less than 50°C. For this type of distillation, the fractionating column is vertically inserted between the flask containing the liquid and condenser.

A fractionating column consists of a long vertical tube through which first the vapour goes up and then is partially condensed which comes down to the flask. This condensed liquid (having higher boiling point component) when flows down through the column gets in contact with the ascending vapours and results in the interchange of heat.

To reach the equilibrium within liquid-vapour system, the vapours get enriched with the more volatile component at the expense of the liquid. And therefore, ascending vapours become richer and more volatile i.e. the lowest boiling component while the descending condensate becomes richer in the highest boiling component. Efficiency of a fractionating column is determined by the extent of separation. Fractionating columns, which may be used in the laboratory, are shown in Figure 2.6 (a-d).

Figure 2.6(a) shows the Vigreux column having moderate efficiency and the most widely used column. It is made by a glass tube with a series of downward slope indentations such that points of each pair of indentations must almost touch each other and form a spiral of glass inside the tube so that the vapour cannot pass directly through the whole length of the column without meeting an indentation.

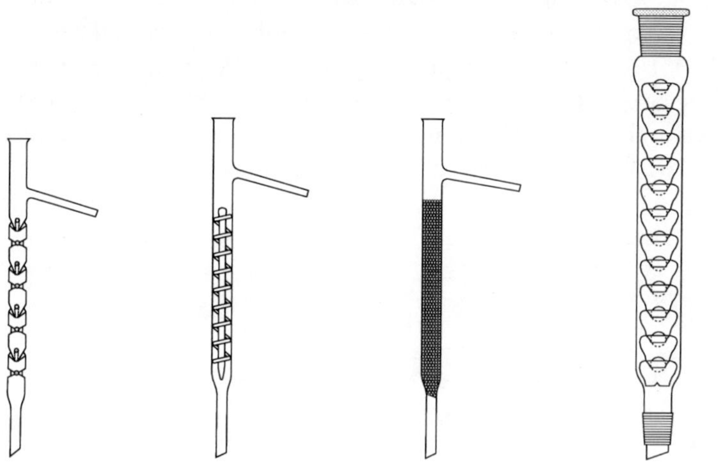

(a) (Vigreux column) (b) (Dufton column) (c) (Hempel column) (d) (Pear-bulb column)

Fig. 2.6 (a-d) Types of fractionating column

While Figure 2.6 (b) shows all-glass Dufton column having efficiency greater than Vigreux column. It is suitable for general use. A glass spiral is fitted tightly in a plain tube around a central tube by grinding the edges of the spiral into the tube. This column has an advantage that a low volume of liquid is retained in the column.

Figure 2.6 (c) shows Hempel column whose efficiency is also greater than the Vigreux column. It is a simple glass tube packed with a suitable packing e.g. 3/16" × 3/16" glass or porcelain rings. It has very high efficiency.

Figure 2.6 (d) shows pear-bulb column, the increase in cooling surface is provided by these pear-shaped bulbs and the efficiency of this column can be increased by increasing the number of bulbs in the column.

The apparatus employed for the fractional distillation is shown in the Figure 2.7. The impure liquid is placed in a round bottommed flask of suitable size, one-third to one-half, pumice stones are added and then the column is fitted absolutely vertical in the position followed by fixing of a water condenser to the side arm and the distillate is collected in the receiver. The thermometer is placed in the position such that its bulb should be just below the level of the side arm. The mixture is heated with a low flame. Care should be taken that the flame should be devoid of draughts so that uniform heat is supplied (use of an air bath, Figure 1.43 gives better results). Initially heating of the liquid should be slow so as to avoid the choking of the column [on fast heating, extra condensation takes place while the column is warming up and so column may be choked with the liquid]. When once the distillation has started, the flame should be adjusted that about one drop of the liquid passes in 2-3 seconds so that an efficient fractionation can be obtained. First, the low boiling point

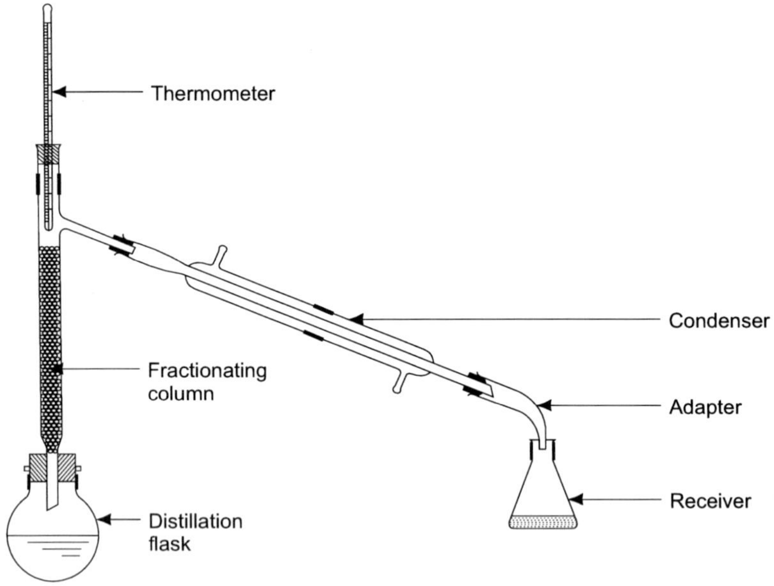

Fig. 2.7 Apparatus for fractional distillation

component will be passed over so when it has passed, distillation should cease. Then heating is slowly increased and a second fraction starts distilling over with a sharp rise in the boiling point. A relative large intermediate fraction may be obtained, if the distillation set-up is inefficient. The distillation should be conducted slowly so as to obtain pure fractions, otherwise the fractionation has to be repeated.

If the boiling point of any of the component exceeds 100°C lagging of the column is necessary, as it will avoid the excessive cooling. Lagging is done by wrapping asbestos cloth or cotton wool around it. Excessive cooling may also be avoided by surrounding the column with a vacuum jacket or an electrically heated jacket.

For small-scale work, the apparatus generally used is shown in Figure 2.8. It consists of a round bottommed flask with a long fractionating sidearm packed with 3/16" × 3/16" glass or porcelain rings.

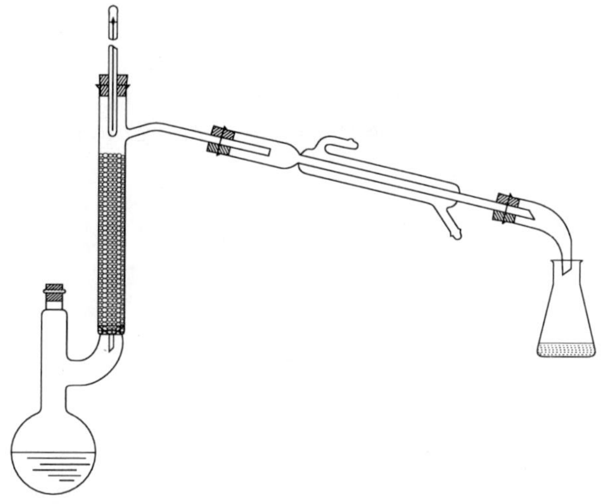

Fig. 2.8 Apparatus for micro-scale fractional distillation

A special receiver may be used to collect the fractions of different boiling ranges of a mixture without interrupting the distillation process, which is shown in Figure 2.9.

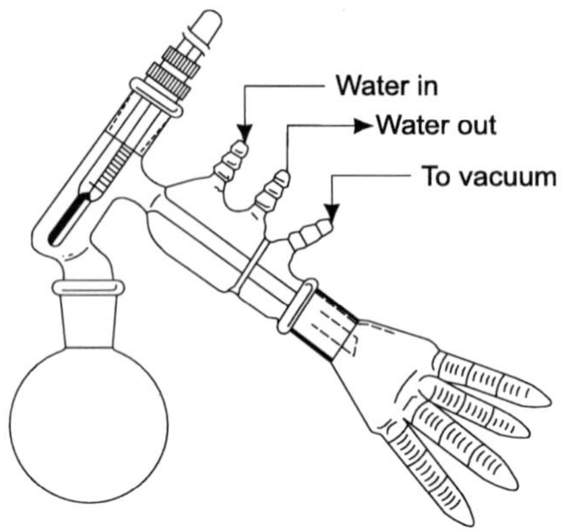

Fig. 2.9 Micro-scale distillation apparatus with a rotating fractional collector

2.4.1 Exercise in Fractional Distillation

(a) Fractional distillation of a mixture of benzene and toluene

The boiling point of benzene is 80°C while that of toluene is 110°C and as the b.pt. difference is 30°C (less than 50°C) and so the mixture can be separated by fractional distillation. It is carried out by using the apparatus shown in Figure 2.7, and here the lagging of the column is not required. For about 40 ml of a mixture containing equal volumes of benzene and toluene, about one and a half hour is needed if distillation is carried out slowly. The flame should have no draughts during the process. Fractions of boiling points (i) 80-85°C, (ii) 85-107°C and (iii) 107-111°C are collected in different receivers. The approximate volumes of each fraction are 19, 2 and 17 ml respectively.

2.5 DISTILLATION IN VACUUM

Some organic compounds decompose before their b.pt. are reached and so they cannot be distilled under atmospheric pressure. In such cases, the compound is distilled under reduced pressure so the boiling point is reduced and hence can be distilled without decomposition e.g. aniline has a boiling point 184°C at 760 mm Hg while it falls to 151°C at 300 mm Hg and to 77°C at 155 mm Hg.

Generally, as the external pressure is reduced by a factor of two, the boiling point of a compound will decrease by about 20-30°C. A nomograph (shown in Fig. 2.10) is useful for estimating the boiling point as a function of pressure. It estimates not only the expected reduced pressure boiling points from the normal boiling points but also the normal boiling points

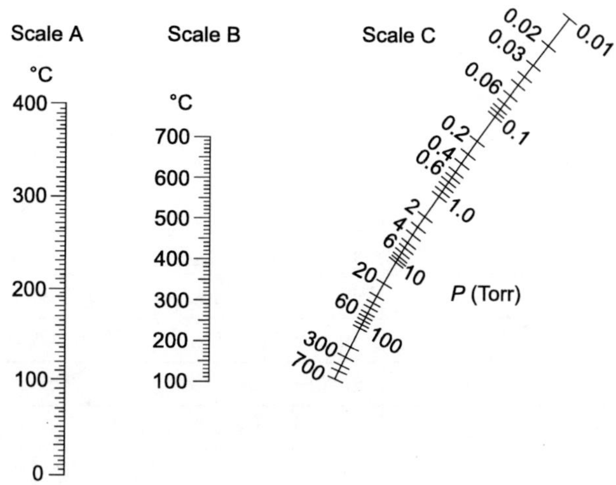

Fig. 2.10 A Namograph for estimating the boiling point as a function of pressure

from observed reduced pressure boiling points. It is applicable for non-associated liquids only. For associated liquids, the variation of boiling point with pressure is 10-20% less than that for non-associated liquids.

The nomograph relates the normal boiling point of a substance (Scale B) to boiling points at reduced pressures (Scales A and C). A line joining points on two scales will intersect the third scale at some point. So, if the values of A and C are known (boiling point A at reduced pressure C), B (normal boiling point) can be estimated and if the values of B are known, boiling point A at a reduced pressure C can be estimated.

Table 2.2 shows the approximate boiling points of some of the substances at reduced pressure.

Apparatus employed for such type of distillation is shown in the Figure 2.11 (a-b).

A specially designed flask, known as Claisen flask is used [As a liquid can bump violently when superheated even under normal distillation process so the possibility of bumping is greatly increased when distillation is carried out under reduced pressure, so to reduce the risk of superheating and hence danger of bumping, Claisen flask is used]. Claisen flask has two necks. The right-hand neck carries a thermometer with the bulb just below the level of the side arm, while the left hand one is carrying a stout capillary tube, whose upper end is closed by a short piece of pressure tubing and a screw clip to regulate the amount of air passing through the capillary. The capillary used here should not be prepared by ordinary glass "quill" tube, as

Table 2.2 Boiling points (°C) at reduced pressures

S.No.	*Name of the Compound*	*Boiling point in °C at reduced pressure in mmHg* $[bp_{\text{red pressure (mmHg)}}]$
1.	Water	$bp_5 1°/bp_{20} 22°/bp_{25} 26°/bp_{760} 100°$
2.	DMF	$bp_{10} 40°/bp_{39} 76°/bp_{760} 153°$
3.	DMSO	$bp_{0.37} 20°/bp_{5.11} 56.6°/bp_{17} 83°/bp_{760} 189°$
4.	Chlorobengene	$bp_5 10°/bp_{25} 39°/bp_{50} 54°/bp_{760} 132°$
5.	Formamide	$bp_{10} 109.5°/bp_{60} 147°/bp_{200} 175.5°/bp_{760} 210.5°$
6.	Glycerol	$bp_{10} 167°/bp_{20} 182°/bp_{30} 192°/bp_{760} 290°$
7.	Quinoline	$bp_{17} 114°/bp_{40} 136.7°/bp_{100} 163.2°/bp_{760} 237.7°$
8.	Ethyl salicylate	$bp_{10} 105°/bp_{25} 124°/bp_{50} 139°/bp_{760} 234°$
9.	Diglyme (Dimethyl ether of diethylene glycol)	$bp_3 20°/bp_{35} 75°/bp_{200} 116°/bp_{760} 162°$

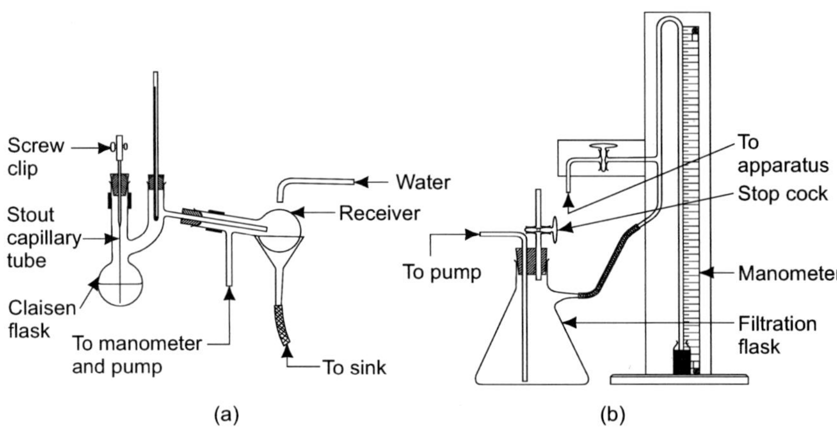

Fig. 2.11 (a-b) Apparatus for vacuum distillation

by this method, capillary obtained will be very fragile and probably may snap during the course of the distillation. It should always be prepared by drawing out a piece of thick-walled capillary tube of 3-5 mm external bore and then the length of the fine drawn-out capillary is adjusted so that it reaches to within 1-2 mm of the bottom of the flask. Then, the side arm of the Claisen flask is fitted to an ordinary distillation flask, acting as a receiver, in such a way that it is well into the bulb of the receiver for the complete condensation of the vapours. Well-fitting and well-bored rubber stoppers and heavy-walled rubber tubing should be used throughout the apparatus. Cooling of the receiver is essential for complete condensation of the vapours and so for liquids having boiling points below 140-150°C, the bulb of the receiver is cooled with a stream of water and the waste water is collected by funnel below, as shown in Figure 2.11 (a) and then carried to the sink with the help of the rubber tubing while for higher-boiling liquids, cooling is usually carried out by dipping the bulb of the receiver in cold water.

The distilling flask (receiver) is then fitted to filter flask, acting as a reservoir to equalize the pressure, by a pressure tubing, which is then connected to a manometer and a water pump as shown in the Figure 2.11 (b). The glass tube, which connects the filter flask to the pump, should touch the bottom of the flask so that any water, which may flow back due to the unequal pressure of water, may be sucked back as the water pressure retains. The larger flow back of water may be checked by opening the stop-cock (connected to the filtration flask) until the original water pressure is restored.

The reduction of pressure is carried out by using a water pump or oil pump. In water pump, the vapour pressure of water at that temperature is the lowest pressure that can be achieved e.g. in winter, the temperature of water is 6-8°C and the vapour pressure of water at this temperature is 6-8 mm, while in summer, when the temperature of water may be 22°C, the vapour pressure of water is 22 mm and therefore, in a water pump, the pressure varies from 7 to 22 mm. For lower pressures, oil pump should be used.

To start the distillation, first the liquid is placed in a Claisen flask (it should not be more than half-filled) and few pumice stones are added to it, the complete apparatus for distillation is fitted as shown in Figure 2.11 (a-b). Reading of barometric pressure is noted down. Now, the flask is heated by an air bath (Figure 1.43) or by a water or an oil bath, by immersing two-thirds of the flask into the bath. The water is supplied to the condenser and then water pump is turned on to attain its maximum capacity, keeping screw clip almost fully closed.* A fine steady stream of air bubbles is passed through the solution by adjusting the screw clip. This will minimise the bumping as the introduction of air prevents the delay in the appearance of the vapour phase and thus superheating. And the volume of air passed in the form of bubbles is so small that there is almost negligible

*If the flask contains traces of volatile solvents, warm the flask slightly and pass a comparatively large volume of air through liquid, so that the volatile solvents get removed down the water pump, otherwise the pressure obtained will have value above the capacity of the pump which can lead to either non-functioning of the pump or leakage in the apparatus. After the removal of all traces of volatile solvent, the screw clip is either almost completely closed or adjusted.

effect of partial pressure on the boiling point when the mercury level in the manometer is constant, the reading is noted. To obtain the value of pressure in the system, the above value is subtracted from atmospheric pressure. The apparatus set-up is considered to be satisfactory if the pressure obtained does not differ by more than 10 mm from the expected value [the temperature of the tap water], while if the pressure is unsatisfactory, the apparatus must be checked for leaks that all glass joints are firmly in position and proper fitting of all the pressure tubings over glass tubings. After achieving the satisfactory pressure, heating of flask is carried out by using a water bath, oil-bath or air bath. When water-bath or air-bath is used, the temperature of the bath should be 20-25°C above the boiling point of the liquid at that pressure, while if an air-bath is used, the temperature is raised slowly until the liquid starts to distil and heating is maintained so that the liquid distils at the rate of 1-2 drops per second. During the distillation, reading of temperature and pressure should be recorded by the thermometer and the manometer respectively. During the distillation, the boiling point of the pure compound should not be raised more than one to two degree, even if the temperature of bath is increased in the end to take off the last drop of the liquid. Sometimes, the liquid distil out first is having the boiling point lower than expected and so in such cases, the heating is continued until the expected temperature is reached and then the liquid is collected in another receiver. To remove the receiver, first the flame is removed and the Claisen flask is allowed to cool slightly by lowering the bath or raising the flask, then vacuum is released gradually by opening the stopcock of the filter flask and screw clip of the filtration flask, the latter will help in preventing the flow of liquid in the capillary.

Now receiver is replaced, again all joints are tightly joined and the whole procedure mentioned above is repeated. When the pressure is having constant value, the heating is started again.

The boiling point at a given pressure may be estimated approximately for the working pressure of a water pump (10-25 mm) by assuming that one mm difference in pressure corresponds to one-degree difference in boiling point. And the more accurate method is by using the expression:

$$\Delta t = 0.00012\ (760 - P)\ (T + 273)$$

where, Δt = correction in the °C to be applied to the observed boiling point T and barometric pressure is P.

For water, alcohols, acids and other associated liquids, the expression used is:

$$\Delta t = 0.00010\ (760 - P)\ (T + 273)$$

2.6 FRACTIONAL DISTILLATION IN VACUUM

If more than one fraction distils, the whole process of simple vacuum distillation has to be stopped after collecting each fraction, as described above and this is the main disadvantage of the simple vacuum distillation. This can be overcome by using a 'Pig' shown in the Figure 2.12 in place of a receiver flask. So various fractions can be collected without disturbing the distillation.

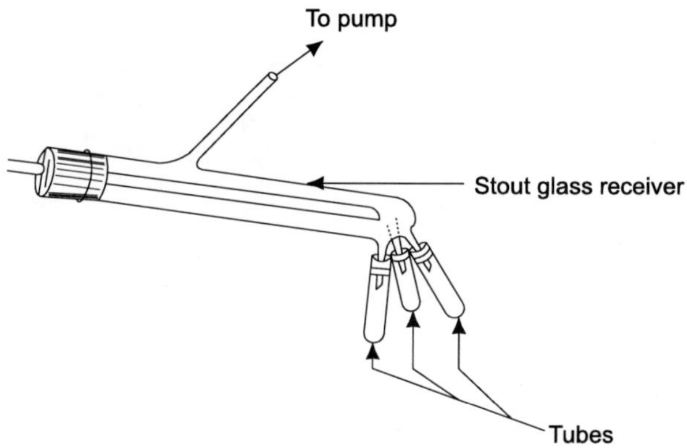

Fig. 2.12 Pig

The 'Pig' contains a glass receiver, which is fitted in place of a receiver flask. It has an outlet tube at the top for the connection to the manometer and pump, and three outlet tubes to which receivers are fitted. So when distillation starts first fraction is collected in one tube and when the second fraction starts coming, then the tube is rotated slightly to bring another outlet in the lowest position so as to collect second fraction in it, similarly third fraction can be collected in third outlet tube. However, this method is unsuitable as if a good vacuum is obtained in the apparatus, rotation of the tube is difficult as then it will be firmly attached while if the rotation is easy around the cork, it means that leakage is there at this point and so varying pressure is recorded by the manometer. The ground glass fitting shown in Figure 2.13 can overcome the leakage problem.

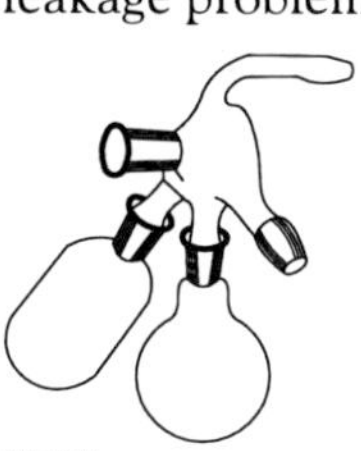

Fig. 2.13 Fractional collector

A very convenient and satisfactory apparatus used for routine work for collecting fractions under reduced pressure is the Perkin Triangle. The complete apparatus for this type of distillation is shown in the Figure 2.14.

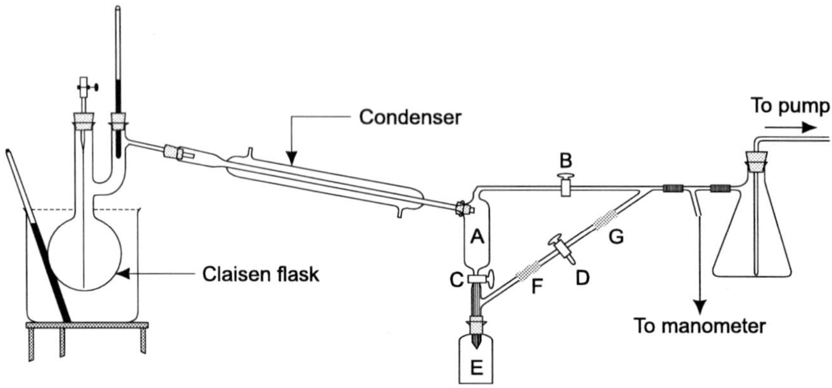

Fig. 2.14 Apparatus of distillation under reduced pressure via Perkin Triangle.

A Claisen flask is fitted (already described) with condenser followed by the Perkin Triangle. The condenser is connected to the receiver 'A' of the triangle, which has two 'two-way' taps B&C and one 'three-way' tap D. These are connected together with glass-to-glass joints by the help of small length rubber tubings, as the Perkin Triangle is very fragile and so these rubber connections help in the replacement of the broken parts at a reasonable cost. The receiver E*, a round bottommed flask or a test tube, is fitted on the double-walled tube from tap C with the help of a rubber stopper. The role of this double-walled tube is that, the inner one is connected from A to receiver E through the tap C while the outer one connects E with the D, a three-way tap, through several small holes present at the lower end below the stopper.

The three-way tap D allows the following connections by suitable rotation : (a) the receiver E with the pump, (b) the receiver E with the atmosphere and tap C is closed so the main system is under 'vacuum' and therefore, the receiver can be changed without interrupting the distillation process and (c) the entrance of air gradually into the whole apparatus (tap C is then open) to attain the atmospheric pressure. The triangle is then

*In vacuum distillation, the flat-bottommed container should not be used to avoid the danger of collapse under reduced pressure.

connected to the manometer, which is then fitted to the pump through the trap to prevent the back flow of the water in the apparatus.

To carry out the fractional distillation under vacuum, the apparatus is assembled completely, the water is turned on in the condenser and a smooth flow of air bubbles is passed through the liquid in the Claisen flask by adjusting the screw clip on the capillary tube. When the vacuum obtained (determined by reading of the manometer) is approximately near to the expected one (differ by not more than 10 mm), the flask may be heated in a suitable bath. When distillation starts, the taps B&C are open and the D, a three-way tap, is so rotated that F & G are connected. Now first fraction is collected from A to receiver E. When second fraction comes, tap C is closed so the fraction collects in the main receiver A. The receiver E is now replaced by another receiver as described earlier. If the pump is having good capacity, on turning D to connect F & G again, the new receiver is evacuated and after a few seconds the pressure is equal throughout the apparatus and now by opening the tap C, second fraction is now run down into the new receiver. While, if the pump is having the small capacity, the sudden entry of the air into the apparatus may cause a slight rise in pressure, and leads to the temporary cessation in the distillation which in turn results in bumping when distillation restarts. This can be avoided by replacing the flame from the bath and to restart heating only when the previous pressure is restored. This whole process of changing the receiver for the collection of various fractions may be repeated as many times as required.

For small volumes of liquids (5-8 ml), the vacuum distillation is carried out conveniently by using the apparatus shown in the Figure 2.15. It consists of a two-necked, pear-shaped flask of 25 ml capacity. The column is charged with Fenske rings. Here the receiver contains six small-rimmed test tubes (75 × 10 mm) fitted into the circular holes of a stainless steel frame, which is supported on a glass tube with the help of the central rod sealed to the inverted stopcock. These tubes can be easily rotated by rotating stopcock, so the different fractions can be collected in the different tubes without stopping the process.

2.7 STEAM DISTILLATION

This method of distillation used to purify the compounds may be solid or liquid, which are water-insoluble and steam volatile [compounds are said to be steam-volatile if they are carried over with the steam when it is passed

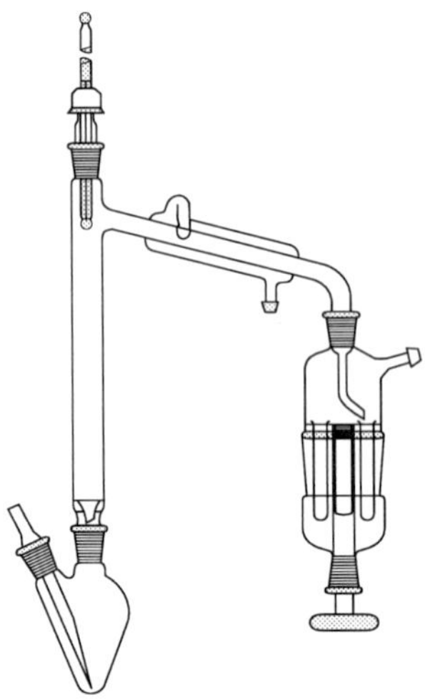

Fig. 2.15 Apparatus for micro scale vacuum distillation

through them]. So the compounds can be purified by passing a current of steam into the mixture of compound and water. The former will distil out with steam and can be easily separated from the distillate (water & compound) as it is insoluble in water. Since steam distillation takes place at a temperature below the boiling point of water,* and so the main advantage is that a compound which decomposes at or near its boiling point can be purified by this method.

So this method is useful for the purification of the organic compound:

(a) which decomposes at or near its boiling point.

(b) by separating the non-volatile by-products in many reactions.

(c) from aqueous mixtures containing dissolved inorganic salts

(d) from those compounds which are not steam-volatile e.g. separation of o-nitro phenol from p-nitro phenol (discussed in exercises on steam distillation).

(e) from steam-volatile by-products e.g. biphenyl and excess of unreacted starting material from the non-volatile triphenylcarbinol (See Sec. 2.7.1).

*As the boiling point, in case of a mixture of two immiscible liquids will be the temperature at which sum of the vapour pressures of these liquids is equal to that of the atmospheric pressure, and its value is lower than the boiling point of the more volatile component and hence the compound distils below its boiling point.

The apparatus for steam distillation is shown in the Figure 2.16. It consists of a steam generator, which is connected to a round bottom flask with the help of a glass tube, which reaches almost the bottom of the flask. This steam-distillation flask is clamped at an angle as shown in the Figure, to prevent the splashing of the solution into the entrance of the outlet tube (which is then connected to the condenser) and thus being blown into the condenser mechanically. The distillate is collected in the receiver flask, which is connected through condenser with the help of a glass adapter.

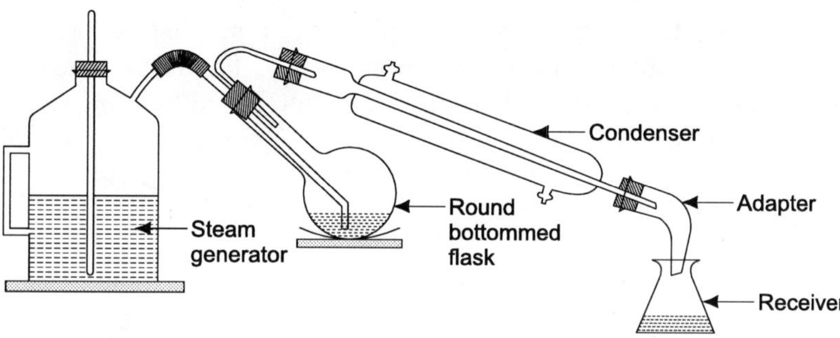

Fig. 2.16 Apparatus for steam distillation

The material (solid with some water or liquid) to be steam-distilled is placed in the flask and a current of steam is passed into the solution and so, the mixture is heated. While the flask itself is heated by a small flame to prevent the rapid accumulation of water. The rate of heating is adjusted so that the vapours are completely condensed by the condenser. The vapours of organic compound and steam are passed over in condenser and get collected in the receiver flask. The steam is passed through the solution until there is no water-insoluble material in the distillate i.e. milky appearance of distillate has just disappeared. If the substance accumulates in the condenser, the water supply to the condenser is turned off for a few minutes until the blockage clears by melting of the solid material and it is now carried over to receiver along with steam. Readmission of water to the hot condenser should be done cautiously. The distillation is stopped (when no milky appearance in the distillation) by first removing the rubber connection between the steam generator and the flask (to prevent the sucking back of the material of the flask to the steam generator). Then the flames are extinguished.

The isolation of the pure organic compound from the steam-distillate depends on the physical state and its solubility in water. The methods of isolation employed are:

(a) If the compound is water insoluble liquid, it will form a separate layer so can be separated by separating funnel.

(b) If compound is water insoluble solid, it will crystallize out in the aqueous distillate, so can be separated by filtration.

(c) If water-soluble organic compound, then the compound can be isolated by solvent extraction [See Sec. 2.15].

To minimize the probability of the splashing of material over into the condenser, the distillation head shown in Figure 2.17 (a) may be used while the probability of leakage problems may also be reduced by using the all-glass head as shown in Figure 2.17(b).

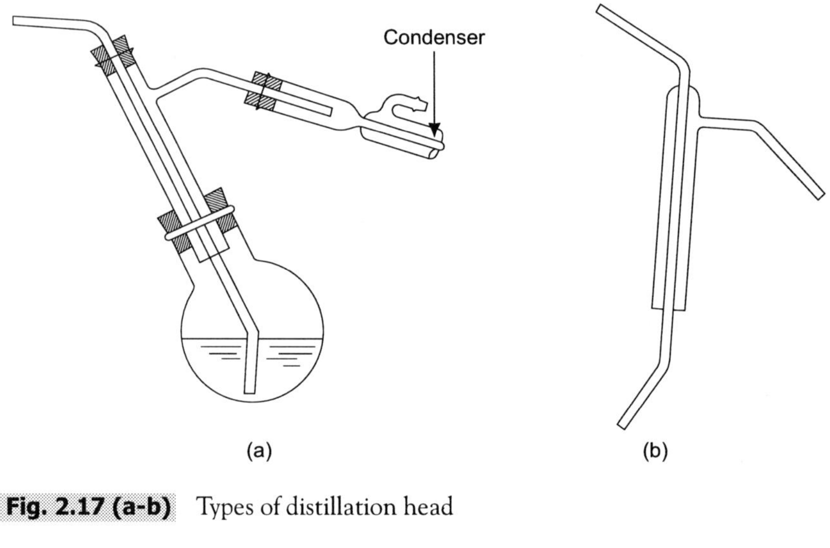

Fig. 2.17 (a-b) Types of distillation head

When the steam distillation is carried out for small quantities, the apparatus, which may be used, is shown in the Figure 2.18. The substance to be distilled is placed in the inner tube, which is dipped in water of the outer flask.

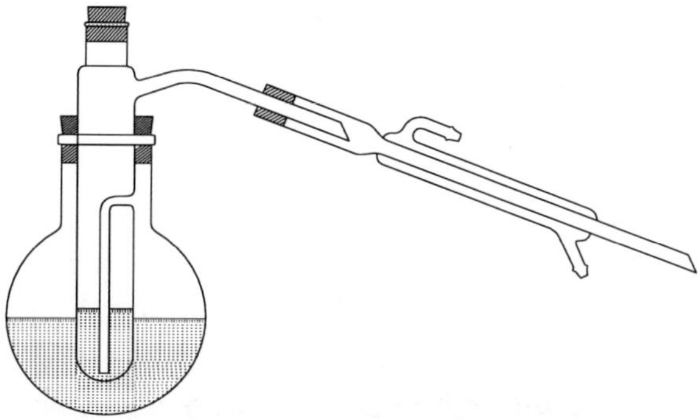

Fig. 2.18 Apparatus for micro scale steam distillation

2.7.1 Exercises on Steam Distillation

(a) Separation of o-and-p-nitro phenol

These two can be separated by steam distillation as the ortho compound contains an intermolecular hydrogen bonding, between oxygen of nitro group and hydrogen of OH group in the ring, while in case of para intermolecular hydrogen bonding is present. This results in high volatility of o-isomer in comparison to p-isomer.

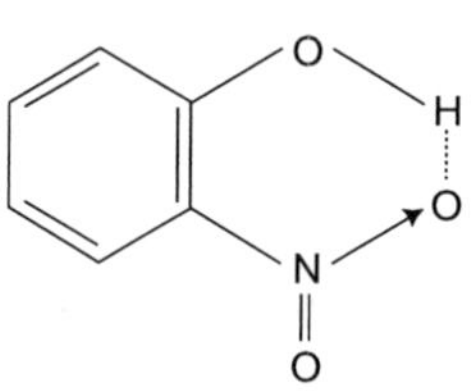

(Intramolecular H-bonding)

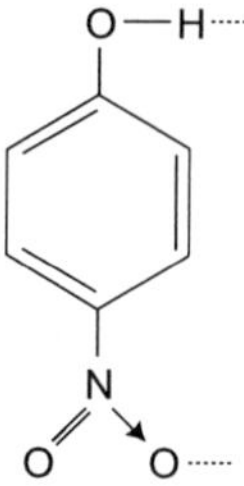

(Intermolecular H-bonding)

These nitro phenols can be prepared by nitration of phenol. So the mixture obtained by the nitration is steam distilled as discussed above until no more o-nitro phenol passes over to the receiver flask. If it gets condensed in the condenser, then the flow of water in the condenser is turned off for a few moments. Distillate is cooled in the ice water for complete solidification of o-nitro phenol. It is filtered at the pump and drained thoroughly. The yield of the dry o-nitro phenol is 7 g, if the nitration is carried out with 20 g of phenol. The m.pt. of pure o-nitro phenol is 46°C.

The p-nitro phenol is obtained by cooling the residue of the flask in ice for 20 minutes. The crude p-nitro phenol is filtered and its purification is carried out by boiling it with 200 ml of 2% HCl along with 1 g of decolourized charcoal for 5-10 minutes. It is filtered through a hot water funnel and filtrate is allowed to cool overnight. The pure p-nitro phenol crystals are filtered and dried. The yield of p-nitro phenol (m.pt. 112°C) is 4 g.

(b) Purification of triphenyl carbinol from the unreacted starting material and the by-products

Triphenyl carbinol, a tertiary alcohol can be prepared by a Grignard reaction, shown below:

$$\underset{\text{(Phenyl Magnesium Bromide)}}{2C_6H_5\,MgBr} + \underset{\text{(Methyl benzoate)}}{C_6H_5COOCH_3} \rightarrow (C_6H_5)_3\,CO\,MgBr + MgBr\,OCH_3$$

$$(C_6H_5)_3\,CO\,MgBr + H_2O \rightarrow \underset{\text{(Triphenyl Carbinol)}}{(C_6H_5)_3COH} + MgBr\,OH$$

$$C_6H_5\,MgBr + C_6H_5\,Br \rightarrow \underset{\text{(biphenyl)}}{C_6H_5 - C_6H_5} + MgBr_2 \text{ (Side reaction)}$$

The residue obtained during this grignard reaction contains triphenyl carbinol, unreacted starting materials and biphenyl, as side product, out of which the former is non-volatile and the latter two are volatile, so can be separated by steam distillation.

To the residue [starting with 6 ml of bromobenzene] 30 ml of water is added and the mixture is placed for steam distillation as discussed above. Steam distillation is carried out until no more oil [unreacted starting material and biphenyl] passes over to receiver. On cooling the residue of the flask, solid separates out, which is filtered at the pump and dried. The triphenyl carbinol so obtained is colourless but crude and so it is crystallized from methylated spirit or benzene. The yield of pure triphenyl carbinol (m.pt. = 162°C) is 4.2 g.

2.8 DISTILLATION WITH SUPERHEATED STEAM

The superheated steam can increase the proportion of higher boiling component in the steam distillate by increasing the vapour pressure of the component in comparison to that of water. The advantage of the use of superheated steam is that there is no need of supplementary heat to the substance as less condensation takes place. The higher temperature of the superheated steam will be obtained by preventing the condensation of steam by surrounding the flask by a liquid bath having the temperature same as that of superheated steam, and this will increase the proportion of higher boiling component in the steam distillate.

The apparatus employed for the distillation in superheated steam is shown in the Figure 2.19.

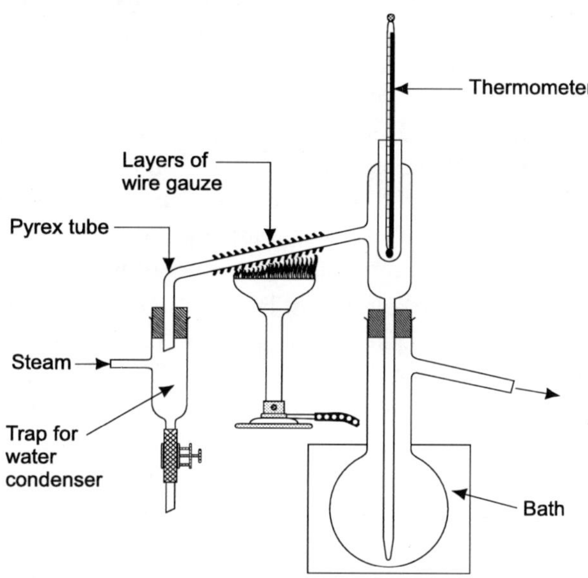

Fig. 2.19 Apparatus for distillation in superheated steam

The flask containing the liquid to be distilled is heated in an oil bath having the temperature approximately same as that of the superheated steam. The Pyrex tube, having diameter (10 mm) is wrapped with a few layers of wire gauze, which is supported by a clamp to prevent sagging of the tube at high temperature. This Pyrex tube is heated with a wing-topped burner. Since the superheater is close to the steam inlet tube, and thus the cooling of the steam enters in the flask, is reduced to a minimum.

2.9 MOLECULAR DISTILLATION

It is also called as High Vacuum distillation or short path distillation. This method is used to distil the compounds having very high boiling points, and which get decomposed at such a high temperature. So to avoid decomposition, they are distilled at pressures in the region of 10^{-5} mm Hg. The main advantage of this method is that the boiling point of the compound is highly reduced [in some cases, by 200-300°C], and therefore can distilled the compounds which decompose at higher temperatures and very sensitive to heat. The distillation of such a high boiling compounds cannot be carried out by using apparatus used for distillation under reduced pressure. To carry out the distillation of above compounds, the main feature of the distillation unit is the short direct path between a heated liquid surface and the cooled condensing area. Since the mean free path of large organic molecules is shorter and therefore the condenser must be quite close to the evaporating surface. A molecular still is a still in which the distance between the evaporating surface and the cold condensing surface is less than the mean free path of the molecules. The temperature of the condenser is maintained comparatively low so that the escaping molecules proceed in a straight path to the condenser.

For small quantities (10-50 mg) of highly viscous high boiling liquids, the vacuum sublimation apparatus (Figure 2.27(a-b)) can be used. The apparatus provides the least hindrance to the flow of vapour from the evaporating surface to the condensing surface. As a liquid sample may contain the dissolved gases or solvents, and so great care must be taken in applying the vacuum. And so first these solvents and gases must be removed. Initially the cold finger is replaced by a stopper. The vacuum is reduced very gradually and the temperature is increased very carefully to avoid excessive frothing and splashing. So initially the vacuum is attained by a water pump and the temperature is raised slowly by immersing the distillation unit in a water bath at a controlled temperature. So the solvent is removed by keeping the froth minimum. When there is no effervescence, the water pump is replaced by an oil pump and the vacuum is reapplied slowly, the heating is continued. When no further solvent is being removed, the stopper is now replaced by the cold finger. The molecular distillation is now started by connecting the apparatus to the high vacuum source i.e. suitable vapour diffusion pump. The complete assembly of the apparatus required for the distillation is shown in Figure 2.20.

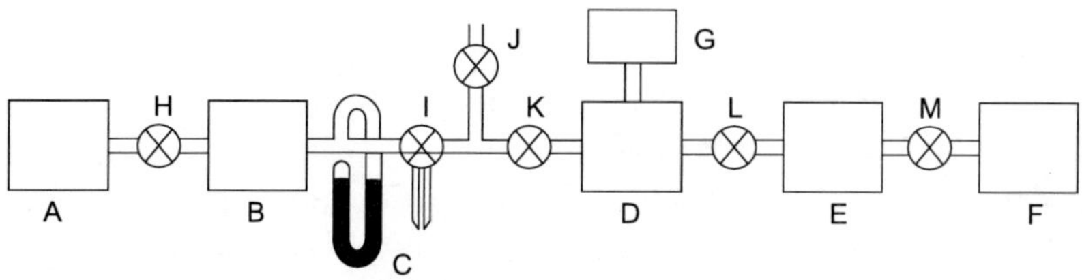

Fig. 2.20 Apparatus assembly for molecular distillation

In the above figure, A is the distillation unit, B is the vapour traps, C is an uncalibrated mercury manometer, D is the vapour diffusion pump, E is a phosphorous acid moisture trap, F is an oil immersion rotatory backing pump and G is the Mcleod Gauze. All these components are connected with wide-bore glass tubing having the minimum number of bends and fitted with joints of the O-ring seal type, or with glass joints sealed with Apiezon Wax W, as required. The H-M are the vacuum taps which site at the points indicating that the assembly can be isolated at these points to facilitate the vacuum control and measurement.

The distillation unit A is attached to the ground glass joint which is fitted to tap H, all the taps are in the closed position. The Dewar flasks, which surround the vapour traps, are filled with suitable coolant e.g. nitrogen water in the cold finger of A and vapour diffusion pump is turned on. The backing pump is also switched and all the taps M, L, K and H are opened in the sequence to attain the pressure with this pump. The pressure in the assembled system is shown by the auxillary manometer C. This can also be used to check the leakages in the assembly by closing the tap M and noting the fluctuations in the mercury level. After reconnecting the backing pump by opening of the tap M, the heat is supplied to the vapour diffusion pump and the pressure is noted down by Mcleod gauze. When the system attains the minimum pressure, the distillation unit A is heated slowly in an oil bath until the misting of the cold finger is observed. The temperature of oil bath is noted and is maintained at this point. The reading on Mcleod Gauze is noted occasionally during the distillation process. When distillation is completed, the heating of the vapour diffusion pump is stopped and the unit A is de-attached by turning off the tap H. The temperature of the fluid of diffusion pump and the unit A is allowed to lower down. Now tap K is closed and a three-way tap M is opened up to connect the backing pump to the atmosphere and then it is switched off.

The vapour traps B are then disconnected for cleaning, drying and reassembly by first opening the pinhole three-way tap I to allow the air to enter in tap B. Now tap H is opened to de-attach unit A. And finally the water supply to the condensers is turned off.

When there is a chance of loss of droplets of condensed liquid from the cold finger, a small collection cup may be attached to it with the help of platinum wire, which is fused, to either side of the glass surfaces shown in Figure 2.21.

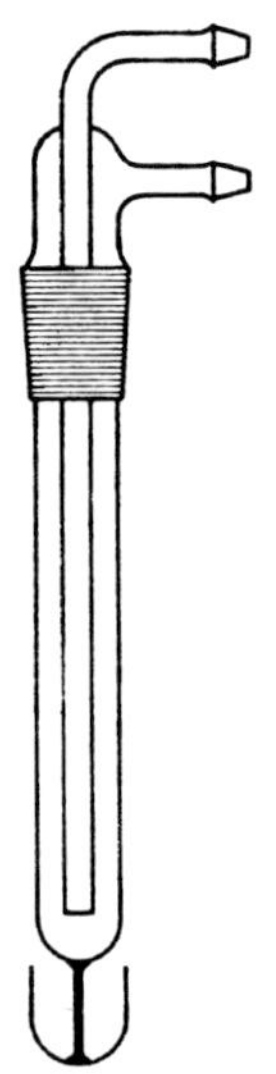

Fig. 2.21 Cold finger with a small collection cup

For small quantities (0.1-2 g) of material, the distillation unit used is shown in the Figure 2.22.

It is formed from Pyrex tubing having 9 mm diameter, the size of the bulbs are made appropriate to the size of the sample. The material to be distilled is added to the distillation bulb in the stages with the help of a capillary tube and so it (material) is diluted with a little solvent. After each addition, solvent is removed, as discussed earlier, and the residue is then distilled by high vacuum distillation. The unit is supported by retort stand

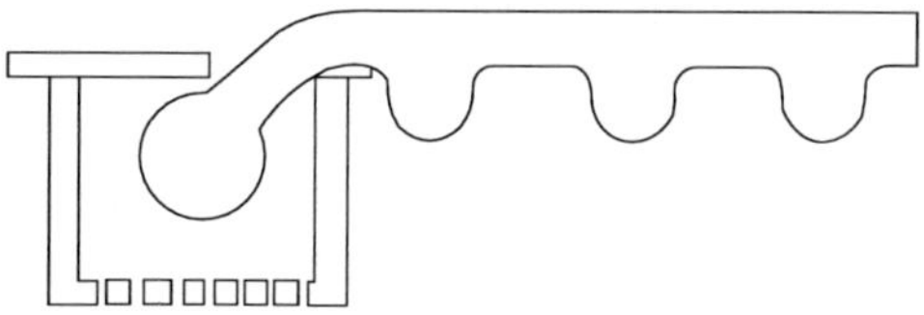

Fig. 2.22 Distillation unit for micro scale apparatus

and clamp in such a way that the receiver tube is sloping downward and the bulb is immersed in an oil or air bath. When the vacuum is attained to a constant value, the heating of the system is started. Liquid distils into the first indentation. The waste of the distilled material is avoided by the further indentations along the tube.

2.10 DISTILLATION OF SOLIDS UNDER REDUCED PRESSURE

The apparatus (Figure 2.14) used for the distillation of liquids under reduced pressure may also be used for solids having comparatively low melting points. In this case, water-cooling of the receiver is not required. While if the solid solidifies in the side arm of the Claisen flask, then the distillation is carried out rapidly and the side arm of the Claisen Flask and the neck of the receiver is warmed with a luminous flame occasionally to overcome this problem.

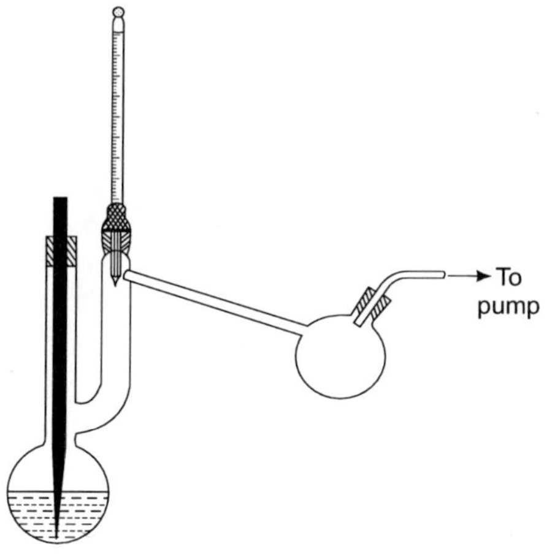

Fig. 2.23 All glass apparatus for distillation of solids under reduced pressure

The all-glass apparatus, shown in the Figure 2.23 is a more satisfactory apparatus and can be used for compounds having melting point as high as 200°C.

This method is mostly used to remove tar, the coloured and non-volatile solid impurities without loss of material. Also, this method of purification of solids is more effective and economical than several crystallizations from a solvent. The purity of the solid can be determined by melting point determination. For purification of solids by this method, the apparatus used is shown in the Figure 2.24. The solid and apparatus must be dry. The flask is heated with a large free flame in rotatory motion round the walls of the flask to avoid bumping which is more frequent in case of heating from the bottom. At the start, if there is frothing, the upper walls and neck of the flask are heating with the flame. After the completion of the distillation, air is allowed to enter the apparatus, and the contents of the receiver are collected by melting it and removed with a bulb pipette. If the distillate is left in the receiver, the side arm may be cut and a new receiver is sealed to the side arm.

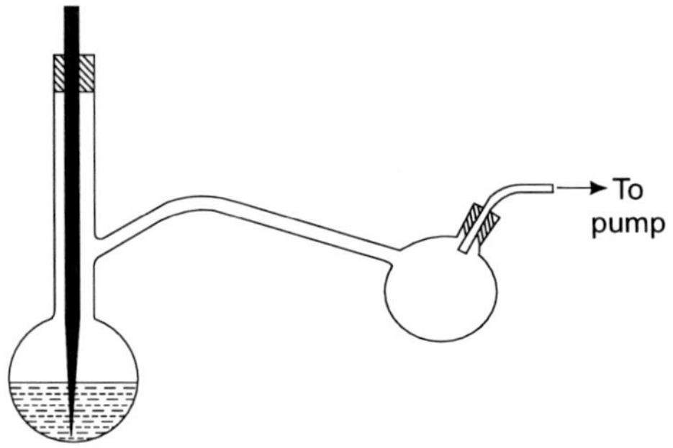

Fig. 2.24 Another type of apparatus for distillation of solids under reduced pressure

2.11 SUBLIMATION

Some of the organic compounds can be purified by the sublimation under normal pressure e.g. naphthalene, anthracene, benzoic acid etc. This method is used for those organic compounds only which directly get converted from vapour to the solid state without passing through the intermediate i.e. liquid state. Such compounds have a high vapour pressure at a temperature below melting point so that the rate of vaporisation from the solid will be fast and also the vapour gets condensed on the cooling surface to give back the solid. This method gives high purity products but with low yield.

The apparatus employed for this process is shown in the Figure 2.25. The dry crude compound is placed in a porcelain dish, which is placed on a wire gauze. The dish is covered with a perforated filter paper [having number of holes, made in an upward direction]. A glass funnel, with a plug of glass wool in the stem, and having the rim smaller than that of the dish, is placed inverted on the filter paper. The dish is heated gently by a small flame, the compound vaporises and passes through the holes, into the funnel, and gets condensed on the upper surface of the filter paper and on the walls of the glass funnel. When most of the compound is vaporized, heating is stopped and the pure compound is collected. One precaution, which should be taken into account, is that heating is adjusted in such a way that the crude material is being vaporised steadily and the funnel does not become more than lukewarm.

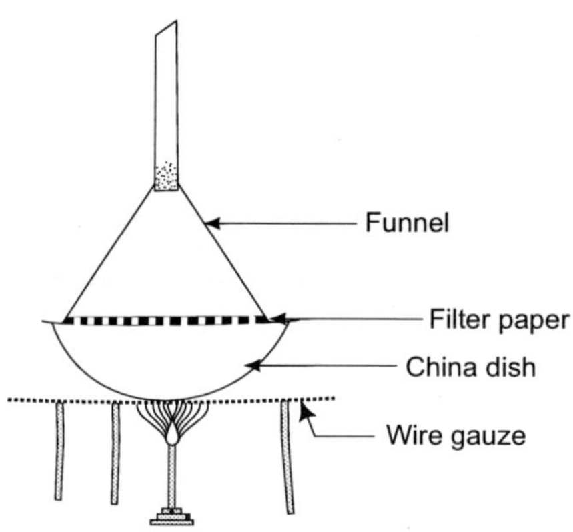

Fig. 2.25 Apparatus for sublimation

Sublimation can also be carried out in a steam of air or inert gas, the apparatus used in this case is shown in the Figure 2.26. The apparatus consists of a two-necked flask with a narrow inlet tube with a stopcock and a wide tube having 12-15 mm diameter, which is fitted to a sintered glass crucible and adopter, which in turn is fitted, in a suction flask. To collect the sublimate, if any, carried by the gas stream, a filter paper is placed on the sintered glass crucible plate.

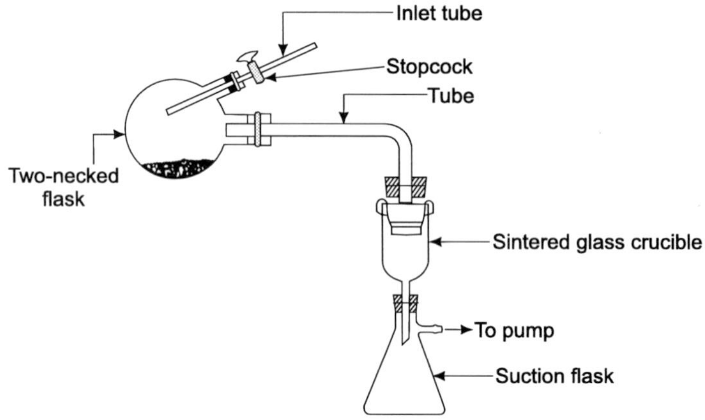

Fig. 2.26 Apparatus for sublimation in an inert atmosphere

2.11.1 Exercise on Sublimation

(a) Purification of anthraquinone by sublimation

Place crude, dry anthraquinone in a china dish, cover it with perforated filter paper and place a cold dry glass funnel, slightly smaller in diameter than the dish, over the filter paper. Heat the dish such that the anthraquinone is vaporised steadily and the funnel does not become more the lukewarm. After a few minutes, the vapours come into the funnel through the holes of the paper and get condense on the upper surface of the filter paper and on the walls of the funnel in long yellow needles. Stop the heating when almost all the anthraquinone is vaporised. Collect the pure anthraquinone. Melting point of anthraquinone is 277°C.

2.12 SUBLIMATION UNDER VACUUM

The yield of the sublimate can be improved to large extent if the sublimation is carried out under reduced pressure and sublimation of the substances having low vapour pressures at their melting points, can also be carried out under these conditions. The one more advantage of this method is to reduce the possibility of thermal degradation, as sublimation is carried out at low temperature under these conditions.

The apparatus employed for the vacuum sublimation is shown in Figure 2.27 (a-b). In figure 2.27 (a), the cold finger is inserted in a larger tube with the help of a rubber stopper and is carrying a disc slightly smaller

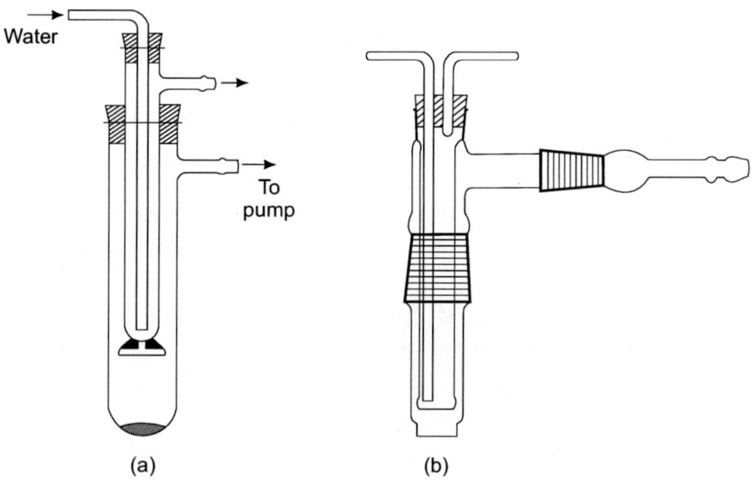

Fig. 2.27 (a-b) Apparatus for vacuum sublimation

in diameter than wider tube. The impure substance is placed in the larger tube and the apparatus is exhausted. Evacuation of the system should be carried out slowly so as to remove the traces of solvent, moisture or air from the substance, which could otherwise cause spattering of the compound on to the cold finger and thus lead to the contamination of the final sublimate. The apparatus is then heated in an oil bath [a direct flame may lead to the decomposition, due to the high temperature], the pure substance collects on the disc and cold surface. One precaution, which must be taken in account during this method, is that the temperature should be allowed to rise slowly till the misting of the cold finger takes place, which is the first indication of sublimation. Even in the final stages, high temperature should not be used to fasten the process.

All glass apparatus shown in the Figure 2.27(b), may also be used. This apparatus can be employed with quantities of a few milligrams upto 1 gram.

2.13 CHROMATOGRAPHY

It is an extremely valuable technique for the separation, isolation, purification and identification of components of a mixture and also to check the purity of the compound [components].

The name chromatography was given to this method because Iswett (1906) first used this method for the separation of coloured substances from plants [Greek, *chromos* means colour].

Purification by fractional crystallization and fractional distillation can be used on a large scale, while they have the disadvantage that they cannot be used on a micro scale. Also fractional crystallization cannot be applied to separate the component whose solubilities are quite similar and fractional distillation cannot be used to separate the components whose boiling points are very close to each other. Similarly sublimation method can be applied on a large scale and also can be used to only limited number of organic compounds. Chromatographic techniques can be used to overcome these disadvantages.

It is based on the general principle of distribution of the components of a mixture between two phases—a stationary phase and a moving phase. Stationary phase can be solid or a liquid supported on a solid, while the moving phase can be a liquid or a gas. If the stationary phase is solid, the basis of separation is adsorption while if the stationary phase is liquid, the basis of separation is partition. Depending upon these phases, the chromatography can be classified into various types:

(1) Liquid-solid chromatography
   (a) Column chromatography
   (b) Thin layer chromatography [TLC]
(2) Liquid-liquid chromatography [Paper chromatography or partition chromatography]
(3) Gas-solid chromatography [GSC]
(4) Gas-liquid chromatography [GLC or Vapour phase chromatography]

2.13.1 Liquid-Solid Chromatography

In this chromatography, stationary phase is solid and moving phase is liquid and hence the basis of separation is adsorption and therefore, this technique can be used for the separation of those compounds, which are having different adsorption coefficients towards a certain solid.

The liquid phase is brought in contact with a finely divided solid phase i.e. adsorbent and the separation of the components of the mixture takes place by the selective adsorption on the adsorbent surface.

The stationary phase, i.e. solid can be packed in the column [column chromatography], or can be coated in the form of thin layer on rectangular glass sheet [TLC].

2.13.1.1 Column chromatography

In this chromatography, solid phase is packed in a cylindrical column, the mixture to be separated is added at the top of the column and the solvent, eluent, is allowed to flow down through the column. A certain fraction of each component of the mixture will be adsorbed by the adsorbent and the remaining will be in the solution. And any molecule will spend part of its time on adsorbent and the rest of the time flowing down with the solvent. The substance having strong adsorption towards the adsorbent will have greater fraction of its molecules adsorbed at any time and therefore any molecule of the substance will spend more time on adsorbent and less time in moving down. While, the substance having weak adsorption will have a smaller fraction of its molecules adsorbed and therefore, will spend less time on adsorbent and more time in moving. Hence, the more strongly adsorbed substance will move down slowly while the less strongly adsorbed substance will move down faster and so comes out in early fractions and the former will come out in the later fractions.

So as the solution of the mixture is added to the column, it penetrates into the column and the components are adsorbed. The fresh solvent is added at the top, when the solution has just completely entered the column. The solvent moves down and re-dissolves the solutes in amounts, which can be calculated by adsorption Law [ratio of amount of solute adsorbed per unit surface area to the concentration of the solute in solution is constant]. So solvent carries solute down the column where it is re-adsorbed, amount of which can also be determined by adsorption law. So as more and more solvent passes through the column, the adsorption and de-adsorption occurs continuously and the solutes gradually move down the column and appear in the form of bands. For solutes having sufficiently different adsorption coefficient, the separation is better. And as discussed earlier, the strongly adsorbed substance will form band above the weakly adsorbed substance. When separation of the band is sufficient enough, the solutes are forced out of the column elution at the lower end and are collected separately. Sometimes, due to the much of time required for the elution, this method to obtain the component from the column is not preferred. In such cases, the surplus solvent of the column is drained off, so that the adsorbent comes out as one piece and the various bands of the column are cut off by laying it down on a sheet of glass and are separately extracted [by adding solvent into it, followed by filtration so that the adsorbent can be removed. Adsorbent is washed 2-3 times with the solvent followed by filtration so as to remove the component completely from the adsorbent. Now the filtrate is distil off to obtain the component as residue].

If the components are coloured, they can be visualized as a series of colour bands. Some compounds are strongly fluorescent under ultraviolet light, so the positions of the bands can be detected by their fluorescence in the U.V. light. But if the components are colourless and non-fluorescent, then the eluent must be collected in successive fractions and the presence of the component is determined by some analytical method. The most common method used for analysis is to distil the eluent. If the residue is a solid, it can be identified by TLC [see Sec 2.13.1.2], melting point, infrared spectrum etc.

2.13.1.1.1 Column

Column is a long narrow glass tube with stopcock. The length and the diameter of the column may vary depending upon the nature of components of a mixture, e.g. if the difference in adsorption coefficient is less, long column is required for the better separation. The apparatus used in the column chromatography is shown in the Figure 2.28. Sometimes

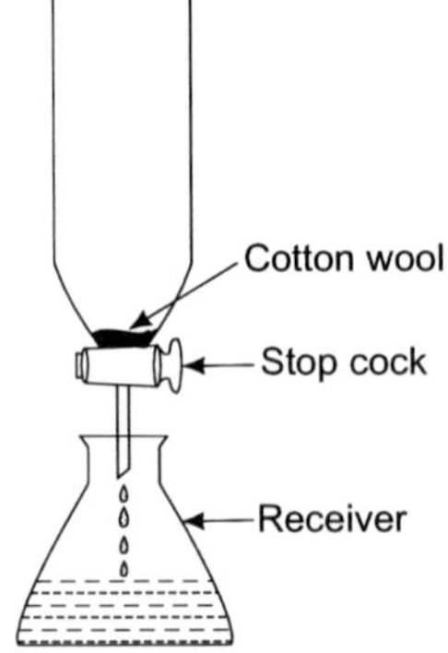

Fig. 2.28 Apparatus for column chromatography

column may be fitted with ground glass joints at its ends so as to allow the attachment of a separatory funnel as solvent reservoir and a Buchner flask via an adaptor with a tap to collect the eluate fractions [Figure 2.29].

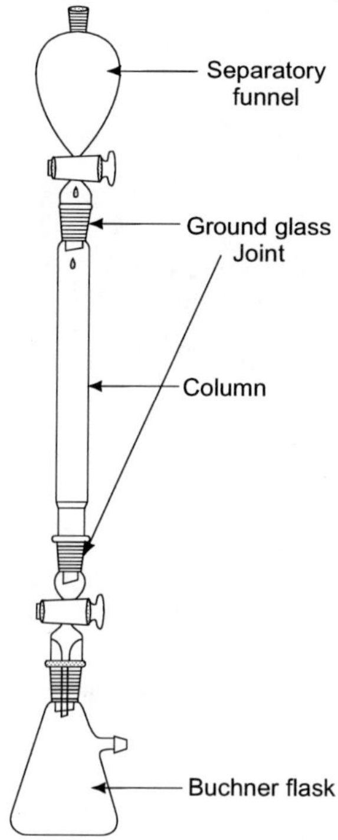

Fig. 2.29 Another type of apparatus for column chromatography

2.13.1.1.2 Adsorbent [Column Packing Material]

Several factors affect the efficiency of a chromatographic separation, adsorbent is one of them. Adsorbent should have following properties:

(a) It should have high and selective adsorption power.

(b) It should have high surface area.

(c) It must be chemically inert.

(d) It must be readily available.

For the separation of any mixture, some adsorbents may have too strong adsorption power that all the components gets adsorbed near the top while some of them will be too weakly adsorbing that all the components move down as fast as the solvent is moving down. Adsorbent should have a maximum selectively towards the components being separated so that there is a large difference between their rates of elution.

Table 2.2 lists a number of adsorbents in order of their adsorptive power. [The order is approximate as it depends on the substance being adsorbed and the eluent, solvent as well.]

Table 2.2 Chromatographic adsorbents

(1) Alumina	Al_2O_3	
(2) Charcoal	C	
(3) Florisil	MgO/SiO_2 (anhydrous)	
(4) Silica gel	$Si\ O_2$	
(5) Lime	CaO	↑ Increasing
(6) Magnesia	MgO	Adsorptive
(7) Magnesium Carbonate	$MgCO_3$	Power
(8) Calcium Phosphate	$Ca_3\ (PO_4)_2$	
(9) Calcium Carbonate	$CaCO_3$	
(10) Potassium Carbonate	K_2CO_3	
(11) Sodium Carbonate	Na_2CO_3	
(12) Talc	MgO/SiO_2 (Anhydrous)	
(13) Sucrose	Carbohydrate ($C_{12}\ H_{22}\ O_{11}$)	
(14) Starch	Carbohydrate [$(C_6H_{10}O_5)_n$]	

The most widely used adsorbent for adsorption chromatography is alumina as it is readily available at relatively low cost. Its particle size [in range 50-200 mm] favours the even packing of adsorbent in the column, reasonable flow of solvent and a rapid equilibrium establishment of adsorbate (solute) between the adsorbent and the liquid phase. It has a wide range of adsorptive power depending upon the extent of water being adsorbed by it. Its adsorptive power can be decreased by adding up to 15% by weight of water. Heavily hydrated alumina can be activated by heating it at 200°C for about three hours. Alumina is obtained in three types:

(a) Basic alumina (pH = 10)

(b) Neutral alumina (pH = 7)

(c) Acidic alumina (pH = 4)

Correct type of alumina must be used to avoid any side reaction, as a particular type will induce a catalytic reaction with a specific functional compound, e.g. the acidic alumina may dehydrate the alcohols (specifically tertiary alcohols) or may isomerise the carbon-carbon double bond while the basic alumina may hydrolyse the esters so in such cases, neutral alumina is preferred.

2.13.1.1.3 Eluting Solvent

Selection of the eluting solvent is also one of the important step in this chromatography. The choice of solvent to introduce the mixture onto the column will depend on the solubility characteristics of the mixture. If the mixture is in the form of a solution, it must be first evaporated till dryness and the residue is dissolved in minimum amount of most suitable non-polar solvent. The concentrated solution should be made so as to form a compact band at the top of the column of adsorbent so that the formation of different bands occurs successfully. It should also show a maximum selectivity in its ability to dissolve or desorbs the substance being separated.

Generally from non-polar solvents, adsorption of the solute (component) on to the adsorbent takes place most readily than from highly polar solvents. And also, frequently the most non-polar solvent used for transferring the mixture to the column and the initial solvent used for the development of the chromatogram is same. Therefore, initially adsorption takes place more rapidly. The development of the chromatogram, if required, can be accelerated by increasing the polarity of the solvent gradually. Table 2.3 lists a number of common solvent in order of increasing adsorbability and eluting power. This order is also approximate as it depends on nature of adsorbent.

A sharp change in solvent composition should be avoided as addition of more polar solvent will result in a large increase in eluting power. And, therefore shifting to more polar solvent is carried out by gradually increasing its concentration i.e. by using successively mixtures of a non-polar and a polar solvent in a ratio of say 90:10, 70:30, 50:50, 30:70, 10:90, 0:100. As water is one of the most strongly adsorbed solvent and therefore, its presence to a very little extent can greatly increase its eluting power and therefore, the solvents used should be completely dry.

The acceptable separation of any mixture by a particular combination of adsorbent and eluting solvent can be determined only by trial. The trial to determine the conditions for a chromatographic separation can be done by using a partially hydrated alumina and a weak eluting power solvent e.g. petroleum ether. Then the solvent of successively higher eluting power is tried until the components move down the column. If the trial adsorbent is

Table 2.3 Eluting solvents for chromatography taking alumina as adsorbent

(1)	Petroleum ether (Hexane; pentane)	
(2)	Cyclohexane	
(3)	Carbon tetrachloride	
(4)	Trichloroethylene	
(5)	Toluene	↑ Decreasing eluting power
(6)	Benzene	
(7)	Dichloromethane	
(8)	Chloroform	
(9)	Ether	
(10)	Ethylacetate	
(11)	Acetone	
(12)	Propanol	
(13)	Ethanol	
(14)	Methanol	
(15)	Water	
(16)	Pyridine	
(17)	Organic acids	

highly adsorbing, it can be replaced by a more completely hydrated alumina or any weaker adsorbent, which if it is too weakly adsorbing, then trial should be done by a less hydrated alumina.

If the substances in the mixture differ greatly in their polarities, they are separated very easily. And so the order in which the components (adsorbate) are eluted from a column depends upon their relative polarities e.g. a hydrocarbon and an alcohol can be separated much easily as alcohol being more polar is adsorbed more strongly on the adsorbent than the hydrocarbon and so hydrocarbon can be eluted with a relatively non-polar solvent and an alcohol can then be eluted with a more polar solvent. Table 2.4 shows the order of ease of elution of the adsorbate.

Table 2.4 The order of ease of elution of the adsorbate

(1)	Saturated hydrocarbons; alkyl halides	
(2)	Unsaturated hydrocarbons, alkenyl halides	
3)	Aromatic hydrocarbons; aryl halides	
(4)	Polyhalogenated hydrocarbon	↑ Decreasing ease of elution
(5)	Ethers	
(6)	Esters	
(7)	Aldehydes and ketones	
(8)	Amines, alcohols and thiols	
(9)	Phenols, carboxylic acid	

2.13.1.1.4 Packing of the Column [Loading of Column]

For the best separation, the column must be packed uniformly without holes, air bubbles, cracks or channels. If the distribution of the adsorbent is uneven, it may lead to the formation of cracks and channels, which may result in the distortion of the adsorption band shapes.

The ratio of amount of adsorbent to the weight of sample to be separated should normally be 50:1. Also, the ratio of the height of the adsorbent in column to its diameter should normally vary between 3:1 to 10:1. More is the ratio, the flow of the solvent may be too low. So for large amount of samples, large-diameter column must be used.

To load the column first a small plug of cotton wool is inserted in the lower end of the column. The adsorbent is mixed with the liquid to form a thin slurry, which is then poured through a funnel into a clean dry column held vertically. The column is tapped gently with a wooden rod so as to settle the adsorbent uniformly. Solvent is removed by the stopcock fitted at the bottom of the column and more slurry is added until the complete required uniform bed of the adsorbent is obtained. Fresh solvent is then flow down the column, until complete setting of the adsorbent is obtained. A disk of filter paper, having diameter just a little smaller than the diameter of the column is dropped on the adsorbent so as to cover the top of the column to protect the surface from disturbance when the sample or the eluting solvent is poured onto it. During the whole process, column loading as well as the separation of bands, the level of the solvent should not fall below the level of adsorbent otherwise the cracking of the bed of adsorbent takes place due to the drying of the column and hence result in the diminished ability to separate the components because of the flow of the solution through these cracks.

A concentrated solution* of the mixture is added to the column and the column is allowed to drain just to the surface of adsorbent. The sample is rinsed down from the walls of the column with a small quantity of solvent, which is again allowed to drain just to the surface of the adsorbent. [Figure 2.30 shows the loaded column]. Now the empty space of the column above the adsorbent is filled with the solvent and a dropping funnel filled with the solvent is attached. The development of

---

*The concentrated solution of the sample to be separated is added so that it gets adsorbed in a minimum layer of adsorbent. And, narrower is the sample bands, overlap is less and hence better is the separation.

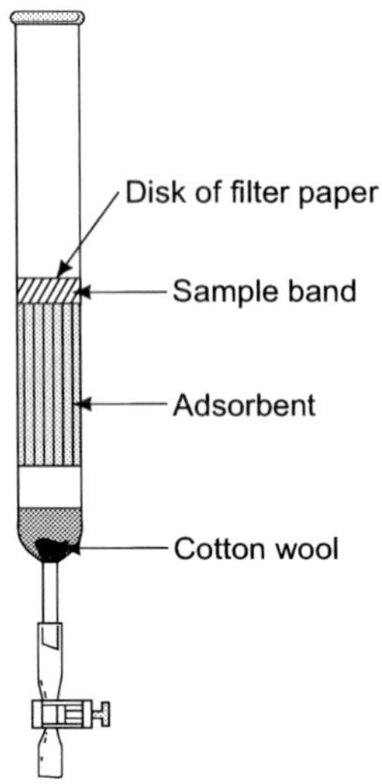

Fig. 2.30 Loaded column

chromatogram, collection, analysis of fractions and recovery of the separated components are as described earlier.

2.13.1.1.5 Advantages and Disadvantages of Column Chromatography

This technique is especially used for purifying small amounts of the compounds. And, the separation by this method is far better than the distillation or recrystallization. But, it is a highly empirical method, and several trials are required to establish the acceptable conditions. Also, it cannot be used to purify the sample in large quantities as even reasonable size columns have a small capacity.

2.13.1.1.6 Exercises on Column Chromatography

(A) Purification of Anthracene

The apparatus employed is shown in Figure 2.29. The column used is 40 cm long and 1.5 to 2 cm in diameter. A 250 ml suction flask is fitted at the bottom of the column and a 50 ml dropping funnel is fitted at the top of the column.

Wash the column with chromic acid mixture followed by distilled water and dry it. Insert a plug of cotton wool at the end of the column. Clamp the column in vertical position. Here alumina is used as an adsorbent. Introduce 3-4 cm column length of alumina and press it by dropping a glass pestle, [which is made from (60 cm × 7-8 mm) solid glass rod and flattened at one end to 10-12 mm] on to it from a height of 5-10 cm several times.

Repeat the process using 3-4 cm portions of alumina each time to have a column of 20 cm adsorbent bed. Insert another small cotton wool plug on the top of the alumina column. Wet the column by adding 25 ml of n-hexane from dropping funnel. Precaution must be taken to cover the column with the solvent during the whole experiment to prevent the formation of holes, cracks, air bubbles or channels. Prepare the solution of technical anthracene by dissolving 50 mg of it in 50 ml of n-hexane and add the solution from the top of the column. n-Hexane is used as an eluting solvent. Develop the chromatogram with n-hexane and examine the progress of the separation with an ultraviolet lamp. After some time, three bands will be obtained:-

(a) a narrow, deep-blue fluorescent band near the top due to carbazole (impurity in anthracene)

(b) a non-fluorescent yellow band in the middle due to naththacene (another impurity in anthracene)

(c) a broad blue violet fluorescent band at the bottom due to anthracene.

Continue the progress of the separation with n-hexane. Change the receiver when fluorescent band (anthracene) begins to come off the column. Discard the elutant collected before the anthracene begins to come off. At this point, elute the anthracene with a 1:1 n-hexane – A.R. benzene mixture. Stop the elution when the middle band (naphthacene) reaches the cotton-wool plug. To obtain the anthracene, evaporate the solvent from the eluate under reduced pressure. Pure anthracene of visibly fluorescent in daylight is obtained. Yield = 30 mg and m.pt. = 215°C.

Clean and dry the column. Discard the alumina.

(B) Separation of o- and p-nitro aniline mixture

Wash the column thoroughly with chromic acid followed by distilled water. Dry it and insert a small plug of cotton wool in the lower end of column and clamp it vertically.

Make a slurry by mixing 100 gm of active alumina with dry benzene and pour it down into the column. Fit a disk of filter paper over the top of the column. And now fit a dropping funnel at the top of the column and a conical flask at the bottom of the column. Allow the benzene to run down the column slowly as the alumina settles in the column. Here also, precaution that column should not be dried throughout the experiment, must be taken into account.

A solution of 0.5 g of mixture in 30 ml of dry benzene is poured into the column slowly through the dropping funnel. Continue the running of pure solvent to develop the chromatogram until the two distinct yellow bands are obtained. a) o-nitro aniline at the bottom & b) p-nitro aniline at the top. Continue this process until each band is collected separately.

Concentrate each of the eluates to about 20 ml by distilling off the benzene in water bath. Evaporate the remaining benzene in an evaporating basin in a fuming cupboard on an electrically heated water bath.

(i) Wash the dry residue of the first eluate with petrol and dry in a desiccator. Pure o-nitro aniline having m.pt. 72°C is obtained.

(ii) Wash the dry residue of the second eluate with benzene and dry in a desiccator. Pure p-nitro aniline having m.pt. 148°C is obtained.

Clean and dry the column. Discard the alumina.

(C) *Separation of 2,4-dinitrophenyl hydrazones*

Clean and dry the column as described in the earlier examples. Insert a cotton wool plug at the lower end of the column. Make the slurry by shaking 15 g of acid washed Kieselguhr and 10 g of bentonite powder in 75 ml of chloroform in a 250 ml conical flask. Pour the slurry slowly into the column. Allow it to settle down by gentle tapping with a wooden rod. Add little more chloroform to rinse down the inside of the column. Insert a small plug of glass wool on the top of the column of adsorbent. Let the solvent level fall down nearly to the glass wool plug and fit the dropping funnel carrying the solution of acetone and benzaldehyde 2,4-dinitro phenyl hydrazones in chloroform* and allow it to run down the column. After 3-4 hours, three visible bands are obtained:

(a) An orange-red band on the top due to unreacted 2,4-dinitrophenylhydrazine

(b) An orange band in the middle due to the acetone 2,4-dinitrophenylhydrazone

(c) A yellow band at the lower end due to benzaldehyde 2,4-dinitrophenylhydrazone

---

*The solution of dinitrophenylhydrazones of acetone and benzaldehyde (crystallized from ethyl acetate) is prepared by warming the 100 mg each of them in 75 ml of chloroform on a water bath until whole of the solid gets dissolved.

Continue the elution with chloroform. First, the yellow band comes out as eluate. When whole of the yellow band comes off, change the receiver and elute the orange band with 1:50 ethanol-chloroform mixture distil of the solvent of each of the eluate separately. First eluate gives 75 mg of the acetone derivative (m.pt.=128°C) and the second eluate gives 98 mg of the benzaldehyde derivative (m.pt. = 236°C). Recrystallize the each derivative from chloroform if melting points are unsatisfactory.

Clean and dry the column, and discard the adsorbent.

2.13.1.2 Thin layer chromatography (TLC)

In TLC, the theoretical basis for the separation of substances is same as that for the column chromatography. The difference between these two methods is the form of adsorbent. In column chromatography, the adsorbent is packed in the column, which, here, in TLC, adsorbent is used in the form of a thin layer (about 0.25 mm thick) on a supporting material, generally a glass sheet (20 × 5 cm is generally used).

The most commonly used adsorbent in TLC are silica gel, alumina, kieselguhr, and cellulose powder. Silica gel is used largely as it has wide applications. It can serve as a medium for the separation of both polar compounds, by partition, as well as, non-polar compounds, by adsorption. These adsorbents are also available with a incorporated fluorescent compound e.g. zinc sulphide to facilitate the detection of components of the mixture by viewing the plates under ultraviolet light. In this method, the glass sheets are coated with these adsorbents by using a binding agent e.g. calcium sulphate to increase the strength of the layer. The prepared thin layer glass sheet is called chromaplate.

2.13.1.2.1 Preparation of Plates

Glass plates are washed carefully with a detergent using a brush to remove the adhere particles followed by the distilled water and are dried in an oven. Precaution should be taken into account in handling of these dried plates that they should only be handled by the edges or the undersurface which is not coated by the adsorbent as any adhere particles on the surface of the plate which is to be coated may lead to the formation of uneven layer and also any greasy contamination may result in the formation of unstable layer.

The assembled apparatus required for coating a glass plate 20 × 5 cm is shown in the Figure 2.31. It consists of a glass plate sheet (20 × 30 cm) to which two glass plates of 20 × 5 cm are placed at the upper and lower ends by means of a cement for glass. In the central depression, the plate to be coated is placed which is held by two uncoated plates one on either side. The required thickness (0.25 mm) of the layer can be adjusted by the spreader.[1]

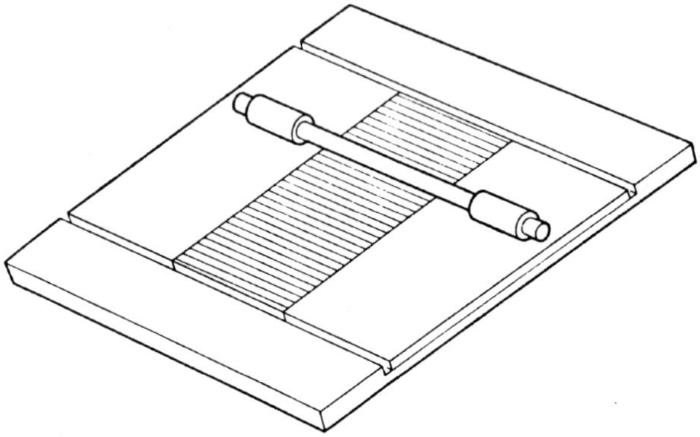

Fig. 2.31 Apparatus for coating a glass sheet

To coat one plate of 20 × 5 cm, 2 g of dry adsorbent is sufficient enough. 2 g of silica gel[2] is mixed with 4 ml of distilled water in a small conical flask and is shaken vigorously for exactly 90 seconds. Slurry can also be prepared by grinding the silica gel G and distilled water in a glass mortar to form a smooth paste. The slurry is poured on the upper end of the central glass plate [Figure 2.31] and is spread evenly over the plate with the help of the spreader. This whole process must be completed in 15 seconds. It is now allowed to stand for 5 minutes to set the layer and then the plate is removed and dried at 110°C for 15 minutes. It is then cooled and kept aside in the atmosphere for half an hour.

Adsorbent coated (with or without fluorescent indicators) TLC plates can also be obtained commercially. As they are prepared carefully under standardised conditions, to get reproducible results, they are preferred to the hand-coated plates due to the elimination of the variation in the factors which can affect R_f values e.g. particle size prior to activation of the plates, thickness of the adsorbent etc. [See sec. 2.13.1.2.5].

---

[1]Spreader is a glass rod 14 cm long and 7.5 mm in diameter with equal lengths (12.5cm) of 2.5 cm. cello-tape

[2]Silica gel G has a particle size of 5-25 μ and contains 13% $CaSO_4$.

2.13.1.2.2 Loading of Plates

To load the chromaplate, first, any excess of adsorbent on the edges and the back of the plate is wiped off. A sharp boundary is formed by drawing a line with a metal scribber, which is parallel with, and 2 mm from the shorter edge of the plate. The excess of adsorbent is removed with the help of the flat side of the spatula and blowing it carefully. The starting position is indicated by making a small mark near the edge of the plate and the starting points must be 1 cm apart.

The solution of the mixture to be separated is prepared by dissolving 3 mg of it in 0.1-0.6 ml of solvent, depending upon its solubility. Rapid evaporation of the solvent is required after applying it to the adsorbent so as to form a spot of small-diameter and hence leads to the better separation of the substances. And, therefore, the solvent must be highly volatile in nature e.g. chloroform, petroleum ether etc.

The solution is spotted on the starting points with the help of a thin capillary tube. The capillary tube is dipped in the solution and then its tip is touched on a piece of filter paper until about 0.5 ml is left in the capillary tube. The solution is applied to the plate by touching its tip on to the adsorbent layer without affecting the surface. It will give the spot of about 0.3 mm in diameter. A very small spot of the solution is preferred as it minimizes the tailing and overlapping of the spots. If the solution is dilute or heavy loading of the sample is required then several small applications in the same place is made by allowing the solvent to evaporate between each addition.

2.13.1.2.3 Development of Chromatogram

The chamber used for the development of the chromatogram can be a beaker covered with a watch glass or a cork stopper bottle [Figure 2.32] for microscope slides while for 20 × 5 cm plate, a cylindrical glass jar is used conveniently [Figure 2.33]. For larger plates 20 × 10 cm or 20 × 20 cm, a rectangular glass tank of suitable dimensions is used [Figure 2.34]. The later can also be used for development of several small sized plates simultaneously.

These jars contain a small quantity of solvent and are lined with filter paper dipping into the solvent. The jar is closed and allowed to stand for few minutes so as to saturate the atmosphere of the jar with solvent vapour. The plate is inserted into the jar by keeping the spots towards the bottom. The level of the solvent in jar is adjusted carefully with the help of a pipette

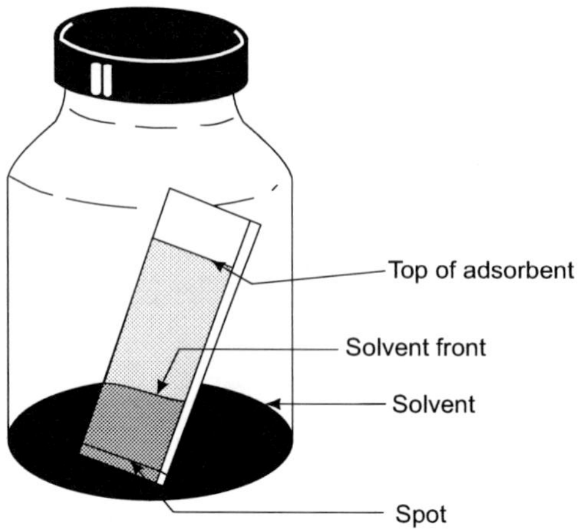

Fig. 2.32 Development of a TLC plate in a corked bottle

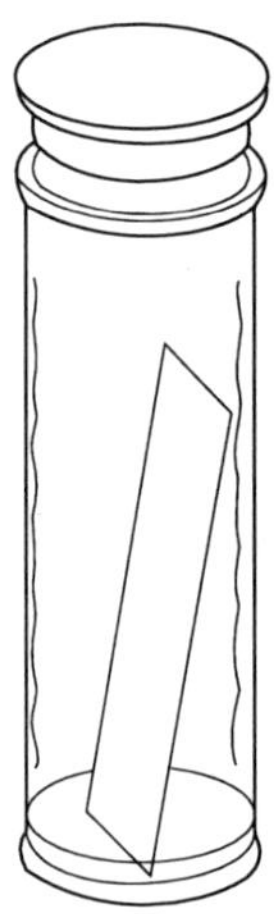

Fig. 2.33 Development of a TLC plate in a glass jar

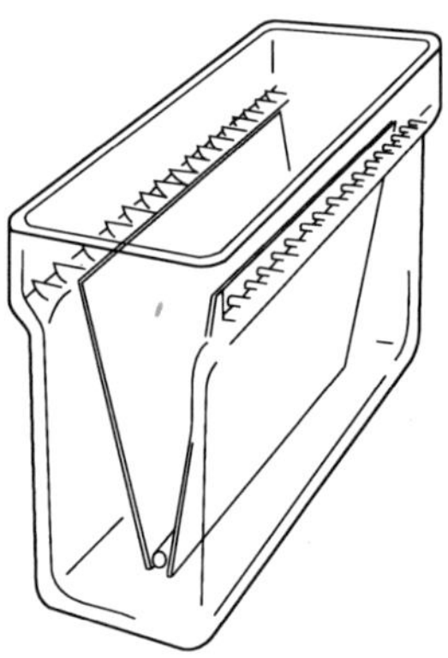

Fig. 2.34 Development of a TLC plate in a rectangular glass tank

so that the lower edge of the adsorbent layer is dipped in the solvent but the spots are far above the level. The jar is closed again and the solvent is allowed to rise on the adsorbent layer by the capillary action. The components of the mixture ascend at different rates depending upon their affinities for the adsorbent. When solvent rises up to 10-12 cm from the starting line, the plate is removed and immediately the position of the solvent front is marked on the adsorbent layer. The time required for the completion of the development of chromatogram varies as it depends on the nature of the adsorbent and the composition of the solvent.

The plate is now dried in a temperature-controlled oven.

2.13.1.2.4 *Visualization of Spots*

If the components are coloured, they can be visualized directly. U.V. fluorescent compounds can be visualized by keeping the plate under an ultraviolet lamp. Non-fluorescent compounds can be visualized due to their fluorescent quenching effect, when they are chromatographed on the adsorbent incorporated with a fluorescent indicator.

A detecting agent used generally for the most of the organic compound is iodine vapour. The dried plate is kept in a closed dry tank containing iodine crystals. Iodine vapours make spots brown. The positions of these spots are marked immediately as the plate has been removed from the iodine jar as iodine vapour evaporates on standing in air so the brown spots disappear.

One more general detecting agent is conc. H_2SO_4 or a solution of 4 ml conc. H_2SO_4 in 100 ml methanol. These detecting agents can be used only when the inorganic adsorbents are used and the components are organic materials. They are sprayed on the dried plate and then the plate is kept for heating in an oven to about 200°C until all the organic materials are turned into dark charred spots.

The detection of colourless compounds can also be carried out by using chemical methods with the help of the suitable spraying agent. These spraying agents are highly selective and may be extremely sensitive, e.g. the ninhydrin reagent* is used for the detection of amino acids, indicators are used for the detection of acids and bases. The advantage of these spraying

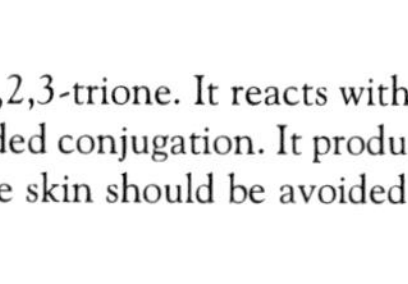

(Ninhydrin)

*Ninhydrin is a 2-hydrate of indane-1,2,3-trione. It reacts with α-amino acids to yield highly coloured products due to the presence of extended conjugation. It produces a long-lasting purple colouration on skin and therefore its contact with the skin should be avoided.

agent is that they can be applied after viewing the plate under ultraviolet light, exposing to the iodine vapour and then allowing the evaporation of iodine. So this extended treatment gives a clearer picture of the composition of the components of a mixture.

The apparatus used for the spray of these agents is called automizer and is shown in the Figure 2.35

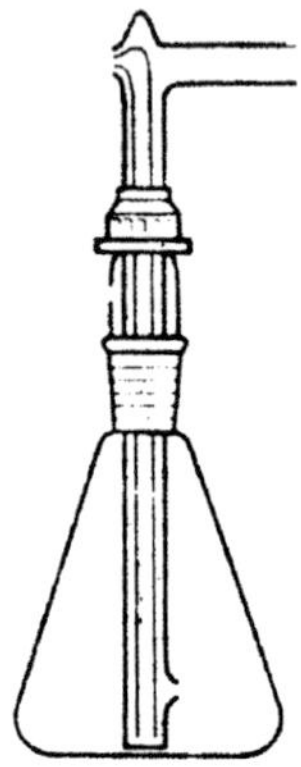

Fig. 2.35 Automizer

2.13.1.2.5 Values R_f

$$R_f = \frac{\text{Distance travelled by substance}}{\text{Distance travelled by solved front}}$$

Under the same experimental condition, for any substance, R_f value is a constant and is characteristic of that substance.

e.g. In the given chromatogram (Fig. 2.36),

R_f value for A, B &C is

$R_{f\,A} = a/d$, $R_{f\,B} = b/d$, $R_{fC} = c/d$

R_f value depends on various factors. These are:

(i) The size of the adsorbent,
(ii) Solvent composition,
(iii) Degree of saturation of the jar atmosphere with solvent vapour,
(iv) Activation and storage conditions of the plates,
(v) Thickness of the adsorbent layer,
(vi) Atmosphere temperature etc.

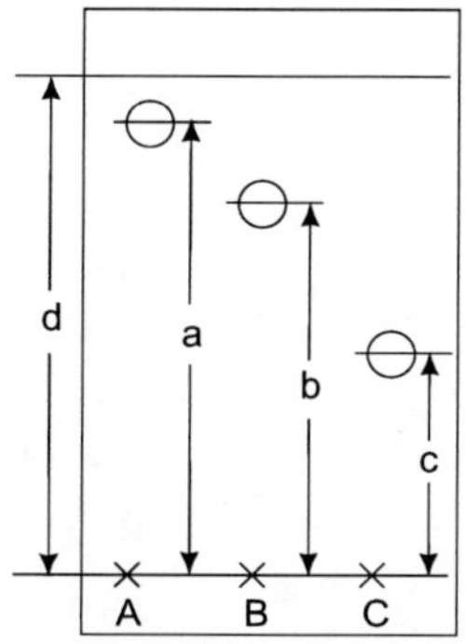

Fig. 2.36 A developed chromatogram

And as even minute changes in one of the above factors, will change the R_f value of the particular substance therefore, experimental R_f values and literature values do not agree very well. And so when reference compounds are available, the unknown and known compound (reference compound) or mixture and reference compound must be run side by side in the same chromatogram.

2.13.1.2.6 Selection of Solvent

For the substances whose chromatographic behaviours are not known, the most effective developing solvent can be achieved by carrying out the preliminary trails using the micro plates in screw-capped bottles. A series of such bottles containing various solvent systems is set up. Micro plates, identically loaded, are developed in these systems separately. They are removed, dried and sprayed with the suitable spraying agent and the mobility of the individual components is noted. When the solvent has ascend completely, the components should be well separated. Solvents, in which, either all the components move near to the solvent front or remain near to the starting point are unsatisfactory. The solvent system which gives a satisfactory chromatogram [well spaced compact spots] is then used for large scale TLC. If no solvent system gives a satisfactory result, the mixture of the solvents should be examined to have the solvent systems of intermediate polarities.

2.13.1.2.7 Precoated TLC Sheets

As precoated glass plates are available commercially [See sec. 2.13.1.2.1], precoated flexible thin-layer chromatographic sheets are also available commercially. The support for the coating, supporting material must have sufficient thickness so as to self-support and must also be flexible. The supporting material may be a solvent-resistant polyester sheet or an aluminium foil, former is used usually.

The sheets are coated with the silica gel incorporated with a small amount of a polyvinyl alcohol binder. It results in a highly porous coating and the solvent penetrates it quickly. These sheets can be activated, if required and can be cut down to the required size with the help of scissors and used as a thin layer glass plate. Advantages of these sheets are that there is the less chance of damage of adsorbent layer and also they can be stored in a notebook etc. after use.

These sheets are also available with the fluorescent indicator.

2.13.1.2.8 Advantages of TLC

It can be used to identify the substance as well as separate them [already discussed in sec. 2.13.1.2]. Paper chromatography [See sec. 2.13.1.3] is used only for the separations on cellulose as other adsorbents cannot form sheets. This limitation is overcome by TLC as large number of adsorbents can be used as they easily form thin layer supported on glass sheets. In TLC, sample spots remain more compact than paper chromatography and therefore very small amount of substances are required for the separation and identification. Larger quantities of the mixture can be separated by using thicker adsorbent layer on large size plates.

Not only these, but TLC can also be used to monitor the progress of the reaction and to assess the success of purification process. In general, the reaction is monitored by the periodic removal of small portions of reaction mixtures and load them directly on TLC plate after their work up.

2.13.1.2.9 Exercises on Thin-Layer Chromatography

(a) Separation of α-amino acids

Separation of DL-α-alanine, L-Leucine and L-lysine hydrochloride.

Dissolve 5 mg of each of the above amino acids separately in 0.33 ml of distilled water (with the help of 1 ml graduated pipette). For Leucine, warm to effect the solution, mix one drop of each amino acid solution to have a mixture of these three amino acids. Dilute the rest of the each solution separately to 1 ml to have solutions of the respective amino acids. Apply spots of each solution on to the TLC plate as already described and allow them to dry in the air.

Prepare the developing solvent by mixing n-propanol (70 ml) with conc. aqueous ammonia (30 ml , specific gravity = 0.88). Line the inside of the jar with the filter paper and allow it to get wet with the developing solvent. Close the jar and allow it to stand for about 10 minutes so that the jar becomes saturated with the solvent vapour. Insert the plate into the jar and add the sufficient developing solvent into it with the help of a pipette, so that the lower edge of the adsorbent layer is dipped into solvent and the spots must be far above the solvent level. Close the jar again and allow the solvent to ascend on to the chromaplate. Remove it when solvent rises up to 10-12 cm and dry it in oven at 100°C for 10 minutes. Violet spots are developed. Mark these spots and evaluate the R_f values from the centre of these spots as discussed in sec. 2.13.1.2.5 [Literature R_f values for alanine, Lysine and Leucine are 0.26, 0.11 and 0.48 respectively].

(b) Separation of 2,4-dinitrophenylhydrazones

Prepare the solutions of 2,4-DNP derivatives of acetone, butan-1-one and hexan-3-one by dissolving 10 mg of each in 0.5 ml of ethyl acetate. Apply the spot of each solution on a flexible silica gel sheet (20 × 5 cm sheets with fluorescent indicator) as already described. Apply the spot of mixture by loading each of the solution one at a time and allow the spot to dry completely after each addition.

Prepare the developing solvent by mixing toluene to light petroleum in 3:1 ratio. Line the inside of the jar with filter paper (b.pt. = 40-60°C) and when the jar becomes saturated with solvent (after 10 minutes), insert the chromaplate into it. Carefully pour more of the developing solvent so that the lower edge is immersed into the solvent but spots are far away from the solvent level. Allow the solvent to ascend on to the sheet. Remove the chromatogram and air-dry it. Calculate directly the R_f values of each component.

(c) Separation of leaf pigments

Prepare the sample solution by crushing the green leaves in a mortar along with few milliliters of ethanol or acetone and twice this volume of petroleum ether or hexane. Filter the mixture. Transfer the filtrate into a separating funnel and wash it with small portions of water thrice. Dry it over anhydrous sodium sulphate.

Here, the developing solvent used is Hexane: acetone, 7:3 (by volume). The developed chromatogram gives the R_f values of carotenes, chlorophyll *a* & *b* and the Xanthophylls.

(d) Separation of a mixture of dye-stuffs

Separation of a yellow dye (Butter Yellow), a red dye (Sudan Red G) and a blue dye (Indophenol Blue).*

Prepare 0.1% of sample solutions of these dyes in benzene. Use Silica gel G as adsorbent and dry benzene as developing solvent. Calculate directly the R_f values of each dye.

(e) Separation of a mixture of polyhydric phenols

Separation of catechol, hydroquinone and phloroglucinol. Prepare sample solutions by dissolving 5 mg of catechol and hydroquinone each in 1 ml of distilled water and 10 mg of phloroglucinol in 1 ml of distilled water. Mix 5 mg of catechol and hydroquinone each and 10 mg of phloroglucinol in 1 ml of distilled water to prepare the sample solution of the mixture.

Prepare developing solvent by mixing 45 ml of benzene, 5 ml of n-butanol and 50 ml of distilled water in a separatory funnel. Shake well for several minutes and separate the upper layer. And Silica gel is used as an adsorbent.

Prepare spraying agent, Phosphomolybedate reagent by dissolving 2 gm of dodecamolybdophosphoric acid in 100 ml of water. A light brown spot corresponds to catechol while a blue spot to hydroquinone. Phloroglucinol gives a faint blue colour which turns brown on keeping for 24 hours by exposing the plate to the vapour of concentrated aqueous ammonia (sp. gr. 0.88) while the colours of hydroquinone and catechol gets intensified on ammonia exposure.

2.13.2 Paper Chromatography

It is a liquid-liquid chromatography as the stationary phase is the water adsorbed by the filter paper via H-bonding (the filter paper is made by highly purified cellulose $(C_6H_{10}O_5)_n$, which is a polyhydroxy compound)

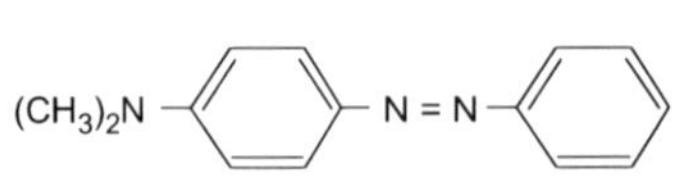

(Butter Yellow)

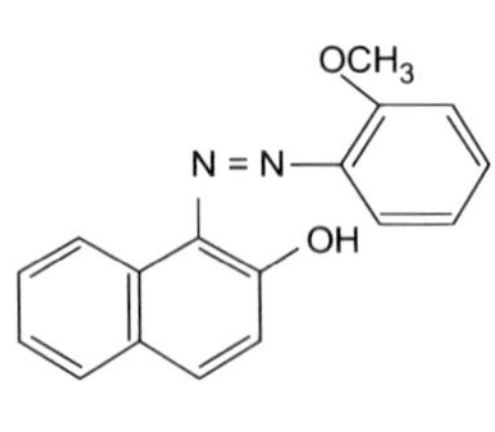

(Sudan Red G)

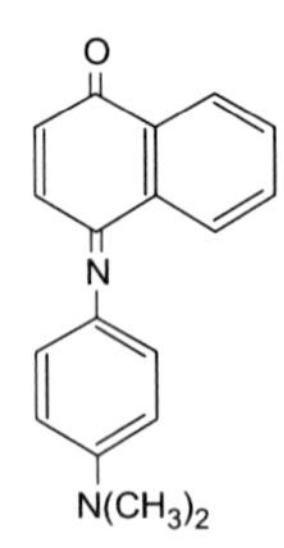

(Indophenol Blue)

and the moving phase is solvent saturated with water. And as the stationary phase is the liquid so the basis of the separation is partition and therefore, this chromatography is a form of partition chromatography.

The process of paper chromatography is very similar to that of thin-layer chromatography except that, the adsorbent on the support is replaced by a strip of paper. Here, also the amount of sample must be very small and therefore, paper chromatography is also used almost entirely as an analytical method to determine the homogeneity of the sample or to determine the composition of a mixture by calculating the R_f values.

Ordinary filter paper can be used for paper-chromatography. Whatman papers, selected grade paper are available commercially. The later ones are more preferred due to a higher degree of uniformity. For general use, Whatman No. 1 paper is used. For heavy loading, a thick grade paper i.e. Whatman No. 3 is used, when speed of development is considered, Whatman No. 4 is most suitable, for good resolution, Whatman No. 20 is required, for non-aqueous phases, modified cellulose papers e.g. silicone treated papers are employed, and for the separation of ionic organic and inorganic substances, ion-exchange cellulose papers are used.

Mobile phase is a mixture of water and one or more organic solvent. Frequently, one of the organic solvent is water-miscible e.g. acetic acid, pyridine to increase the amount of water in the moving phase. Many of the developing solvents form two-phase systems, the organic phase which is saturated with water is used as moving phase, while the aqueous layer is used to saturate the atmosphere of the jar in which the chromatogram has to be developed, e.g. in the paper chromatography of the amino acids (discussed later in the same section).

2.13.2.1 Application of sample

The filter paper strips 12.5 × 17.5 cm from the sheets of Whatman No.1 paper is used to develop the chromatogram. The paper strip should be handled cleanly. A starting line is drawn by the pencil [draw line and mark lightly without affecting the paper] about 2-2.5 cm from the bottom of the short edge and spots are marked apart symmetrically on the line as shown in the Figure 2.37. Apply the solutions of the substances to be chromatographed on these spots with the help of a thin capillary tube to form a circular zone of about 4-5 mm in diameter. Capillary tube can be washed thoroughly and can be reused after drying, fresh capillary tube can

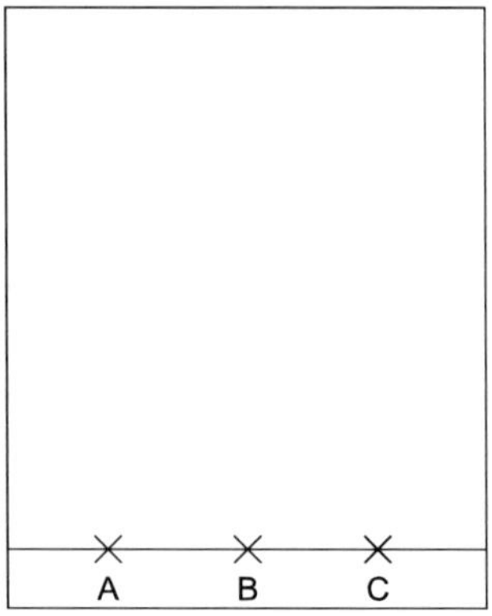

Fig. 2.37 A chromatogram

also be used for each solution. The paper strip is dried in the air. If the solution is dilute, large amount can be applied by making several small applications, and allowing the previous spot to dry before putting the next one.

2.13.2.2 Chromatogram development

After drying the paper, it is placed in a suitable container where it is moistened with moving phase by ascending technique (i.e. moving phase is moving upwards by capillary action) or descending technique (moving downwards by gravity).

(a) Ascending Technique

In ascending technique, the apparatus used are shown in Figure 2.38 (a-b). The paper is placed vertically in a closed jar without touching its sides as shown in the figures. As in TLC, here also, the precaution that the paper edge with starting line at bottom should be dipped into the solvent while spots must be far above the solvent level.

(b) Descending Technique

In descending technique, the apparatus used is shown in Figure 2.39. It consists of a special designed jar having a trough sufficiently supported at the top of the jar for the moving phase. The paper is held within the trough with the help of a glass rod whose one of the end is angled for the suitable handling. Another glass rod is placed close to the trough edge as shown in the above Figure 2.39 to pass the paper over it to protect the paper by the glass edge of the trough and for regular and evenly flow of solvent down the

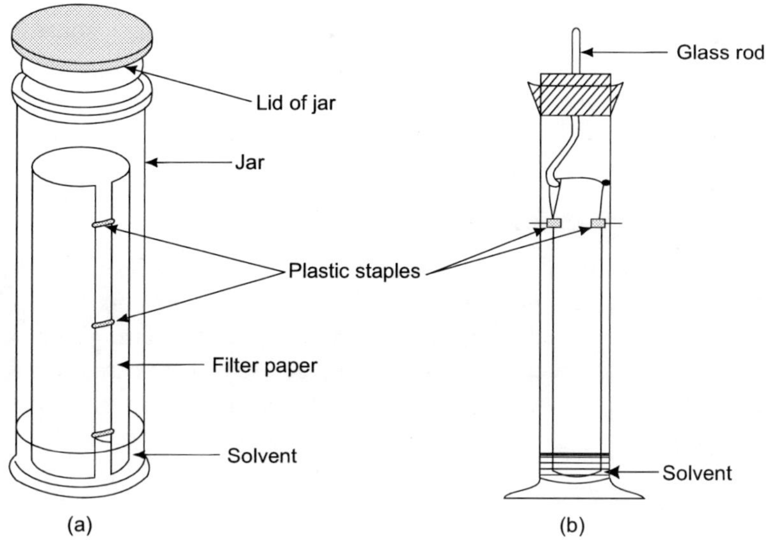

Fig. 2.38 (a-b) Apparatus for development of chromatogram in ascending technique

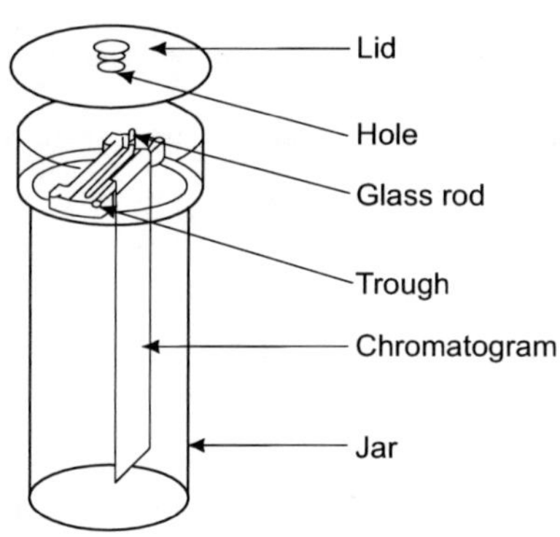

Fig. 2.39 Apparatus for development of chromatogram in descending technique

paper. The two-phase solvent system may be used in this technique, by placing the aqueous phase in the bottom of the jar and when the atmosphere becomes saturated with it, the paper to be chromatographed is then supported in the empty trough. The jar is then closed again and the paper is allowed to stand for 1-2 hours to get equilibrate with the atmosphere of the jar. The upper layer (organic layer) of the solvent system is added into the trough through the hole in the cover. The solvent is allowed to move down for 12-13 hours. Then the cover of the jar is removed and the paper is cut down along the supporting rod and is removed from the jar by lifting the supporting rod. The solvent front is marked with a pencil and paper is dried.

2.13.2.3 Visualization of spots

If the substances are coloured, the spots can be visualized directly. Sometimes, ultraviolet light can be used to visualize the spots. Spraying agents can also be used to locate these spots. Disadvantage of this technique is that the specific reagents should be used which cannot attack the paper. This can be overcome by TLC and so corrosive reagents can be used for spraying.

2.13.2.4 R_f values

As in TLC, here also, R_f values are quite sensitive to the exact experimental conditions. For the comparison of experimental R_f values of unknown substance with the most probable known substances, samples of all these substances should run side by side on the same strip of the paper.

2.13.2.5 Exercises on paper chromatography

(a) Separation of amino acids by descending technique

Separation of glycine, DL-α-alanine and L-Leucine.

Dissolve 5 mg of each of the amino acids separately in 1 ml of distilled water. Prepare developing solvent by mixing 40 ml of n-butanol, 10 ml of glacial acetic acid and 50 ml of water in a separatory funnel. Allow the two-phase mixture to settle, separate the two layers. Use fresh solvent as it undergoes esterification on standing.

Use the container for this technique as shown in Figure 2.38(b). Cut Whatman paper No. 1 in 50 × 10 cm strips. Touch the strips as minimum as possible and handle them by the edges only. Mark a starting line with pencil lightly 2.5 cm from one end. Then mark the points equidistant from each other with the pencil. Apply spots of each amino acid solution on these points carefully with the help of a capillary tube [diameter of these spots should be 4-5 mm].

Add the aqueous (lower) layer of the developing solvent in the bottom of the jar. Locate the paper in the trough vertically and cover with the lid.

Allow the system to become saturated with the solvent vapour by standing it for 1-2 hours then add organic (upper) layer into the trough through the hole in the cover. Allow the development to proceed for 12-13 hours. Remove the cover of the jar, cut the paper along the supporting rod and remove it from the jar. Mark the solvent front with the pencil and dry it in oven at 110°C for about 5 minutes. Spray the paper with ninhydrin [0.2% solution in 2-butanol] and again put it in oven for 5-10 minutes, calculate the R_f values of these purple coloured spots corresponding to the above amino acids.

(b) Separation of dicarboxylic acids by ascending technique

Separation of oxalic, malonic, succinic and adipic acids.

Sample Solution: 0.2 g of each acid in 100 ml of water.

Developing solvent: 6 ml of conc. aqueous ammonia diluted with water which is added to a mixture of 160 ml of ethanol and 30 ml of water.

Spraying agent: Ethanolic bromocresol green (100 mg in 100 ml of ethanol) followed by 3% aqueous lead acetate solution.

Calculate the R_f value of each dicarboxylic acid.

(c) Separation of polyhydric phenol by descending technique

Separation of catechol, resorcinol, hydroquinone and phloroglucinol.

Sample solution: 10 mg of each separately in 1 ml of water

Developing solvent: Mix 95 ml of benzene, 5 ml of n-butanol and 100 ml of water in a separatory funnel. Shake it for several minutes. Allow the two-phase system to settle and separate the two layers.

Spraying agent: Phosphomolybdic acid reagent [by dissolving 2 g of dodecamolybdophosphoric acid in 100 ml water]followed by exposing the chromatogram to the vapour of conc. aqueous ammonia (sp. gr. 0.88).

Procedure is similar to that discussed in example(a) [separation of amino acids].

(d) Separation of monosaccharides by the descending technique

Separation of D-xylose, D-arabinose, D-glucose, D-galactose and D-fructose.

Sample solution: 30 mg of each in 0.25 ml of water

Developing solvent: 40 ml of butan-1-ol, 50 ml of acetone and 10 ml of water.

Spraying agent: Ethanolic naphthoresorcinol [0.2% of napthoresorcinol in ethanol containing 10% v/v of orthophosphoric acid].

Procedure is same as in example(a). In case of monosaccharides, R_f values are low and therefore the solvent is allowed to move down and drip from the paper end. And as there is no solvent front position, the D-glucose is taken as a standard and the chromatographic behaviour of the individual monosaccharide is compared with it. This gives a new constant, R_g.

$$R_g = \frac{\text{distance travelled by monosaccharide}}{\text{distance travelled by D - glucose}}$$

(e) Separation of Anthranilic acid and N-Methyl anthranilic acid sample solution

20 mg of pure anthranilic acid in 20 ml of methanol and 10 mg of pure Methylanthranilic acid in 20 ml of methanol.

Developing solvent: 10 ml of conc. aqueous ammonia (sp. gr. 0.88) in 40 ml of redistilled n-butanol.

Spraying agent: These acids give intense blue fluorescent zones in ultraviolet light. Ethanolic ferric chloride solution can be used as spraying agent:- N-methylanthranilic acid gives a purple brown spot and anthranilic acid gives a very faint pink colouration.

2.13.3 Gas Liquid Chromatography (GLC)

It is a rapid and convenient technique to analyse the composition of the mixtures of sufficiently volatile organic compounds. In this chromatography, stationary phase is a high boiling liquid supported on a suitable granulated solid medium, and the mobile phase is gaseous phase [mixture of components and an inert gas]. Here, the separation is carried out by partition of substances between these two phases.

Large number of stationary phases are available commercially. Some of the examples are:

(a) 

(a long chain hydrocarbon)

They separate molecules in order of their boiling points.

(b) 

(Polythene glycol)

It is good for polar compounds e.g. alcohols, ketones, esters etc. It is better than the hydrocarbons.

(c) 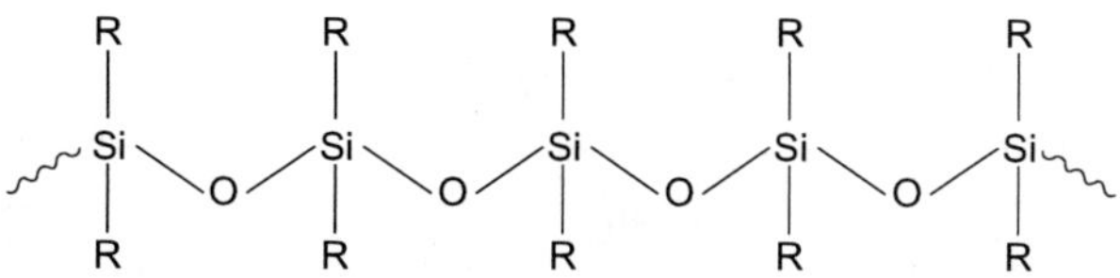

(Silicone oil)

These are generally useful and separate the molecules according to their volatility.

The solid supporting material generally used is Kieselguhr or firebrick, which are available commercially in various particle sizes.

The apparatus required for GLC is more complex and expensive than that for the other chromatographic methods. The principal features of the apparatus employed for GLC is shown in the Figure 2.40.

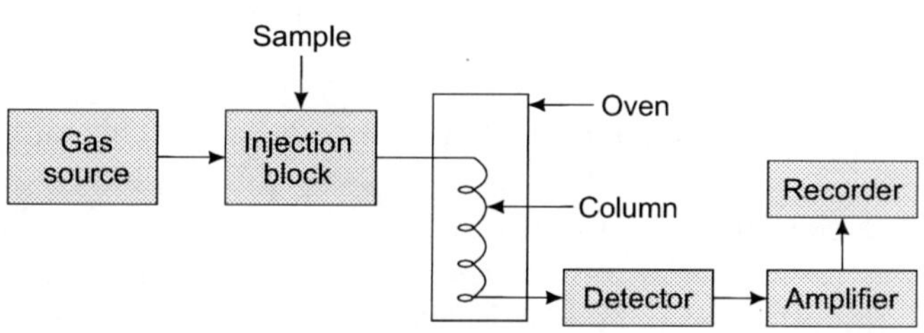

Fig. 2.40 Principal features of apparatus for GLC

It consists of a source of carrier gas, an injection block for the introduction of the sample, a column (made up of metal or glass, 2-3 m length and 2-4 mm internal diameter and is in the form of a circular spiral) in which the supporting medium along with the stationary phase is packed. The components of the sample are carried down the column where they are separated depending upon their partition coefficient between the stationary and mobile phase and comes into a detector. The detector senses the presence of the material in the carrier gas. The signal from the detector is suitably amplified by an amplifier and the recorder records the signal as a function of time.

Liquid samples are injected directly whereas solid samples are dissolved in a volatile solvent, e.g. ether, before injection. A very small amount (0.001-0.005 mL) of a sample is required which is drawn into the syringe without any air bubbles. If the air bubbles appear in the syringe, they can be removed by either dipping the needle into the sample and pumping the plunger for 3-4 times or holding the syringe with the needle up and tapping the barrel with the finger until the bubbles rise to the needle end and then pushing the plunger in to force the bubbles out.

The recorder is started before the injection of the sample. The sample is injected by inserting the needle of the syringe into the rubber septum and pulling the needle straight back out after sliding in the plunger.

A chromatogram may take one minute to half an hour. Recorder is turned off when the chromatogram is complete. If peaks are two small or pen goes off the scale, the chromatogram may be obtained again by changing the sensitivity setting.

The area under the peak should be a measure of the amount of the given compound present in the sample. Higher is the concentration of a given compound in the carrier gas, the larger is the amount of it that comes off the column.

The qualitative and quantitative assessment for the composition of the mixture can be carried out by the examination of the graph produced by recorder. No. of peaks in the chromatogram indicates the number of components present in the given mixture. And the percentage of each component can be calculated by measuring the area under each peak* and taking this as a proportion of the total area. And for a mixture of two substances, the relative amounts of the substances can be calculated by measuring the ratio of the areas of the two peaks.

If the peaks are not resolved completely, but there is somewhat overlapping among them, estimation of the area will be difficult and in such cases, resolution can be increased by using a longer or narrower column, smaller sample size and lower operating temperature.

The identification of the components of the mixture can be made by first comparing the retention times with that of the pure components. And it can be confirmed by adding each of the suspected components in the

---

*Area under each peak can be determined by:

(a) Counting squares

(b) Measuring the peak height from the extrapolated base line and width at half-height. The product of the two is the area.

(c) Tracing each peak carefully on another paper and weighing the area enclosed by cutting these trace peaks.

$$\% \text{ of the component} = \frac{\text{Weight of the component}}{\text{Total weight of all the components}} \times 100$$

mixture carefully and co-chromatograph the resultant mixture. The addition of the same component in the mixture will enhance the appropriate peak, and therefore an enhancement in the peak will confirm the presence of the added component in the mixture.

So GLC method is used not only to estimate approximately the purity of the compound by comparing the area of the main peak (peak corresponding to the main compound) to the area of other peaks, but is also used to determine the presence of a certain compound in a mixture.

2.13.4 Ion-exchange Chromatography

This technique can be employed for the separation of organic compounds on a semi-micro or micro scale. It is a special example of liquid solid chromatography in which strong ionic attractions are present in contrast to weak polar adsorption forces in an ordinary liquid-solid chromatography. These ion-exchange materials are either cation exchange resins or anion exchange resins.

Ion-exchange materials are usually resins having acidic or basic groups. They are available commercially and can be made by co-polymerisation of the styrene and divinylbenzene. The lattice structure is formed by the cross linkages between the chains of polymerised styrene via divinylbenzene. The suitable acidic or basic group required for the exchange, can be incorporated either before or after the condensation.

(a) Cation-exchange resins

These resins are capable of exchanging the hydrogen ions. They may contain sulphonic acid groups $[-(SO_3)^-H^+]$ as their active principle groups. Such polymer is called as strongly acidic cation resins. Instead of sulphonic acid groups, carboxylic acid groups $[-(COO)^- H^+]$ may also be present. Such resins are called weakly acidic cation resins.

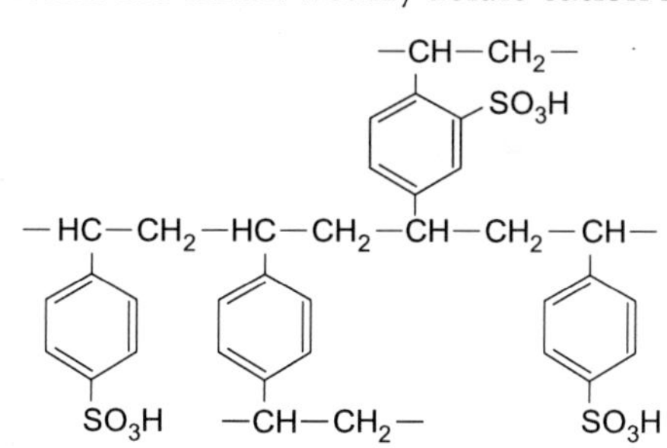

Sulphonated exchanger
(Strongly acidic cation resin)

Both the above type of resins are insoluble acids of extremely high molecular weight.

(b) Anion-exchange resins

These resins are having the structures similar to that of the cation-exchange resins but differ in their capability of exchanging with anions in the surrounding medium. It may contain quaternary ammonium groups $[-(NMe_3)^+X^-$ where X=Cl or Br] as the active exchange principle group and are known as strongly basic anion exchange resins.

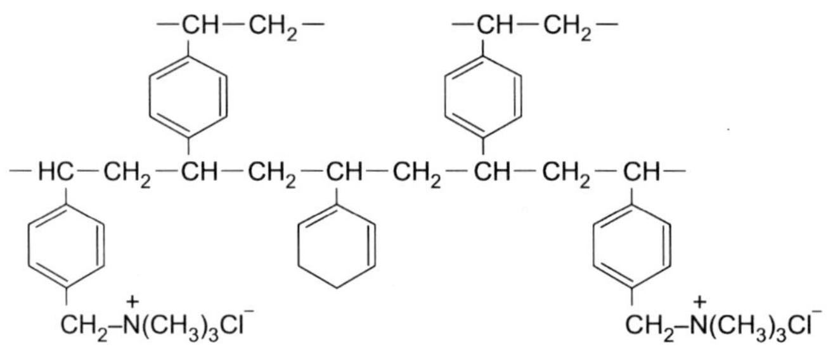

The resins which contain amino groups (e.g. NMe_2) in place of quaternary ammonium groups are called weakly basic anion exchange resins.

2.13.4.1 Applications of ion-exchange chromatography

Ion exchange resins are suitable for micro and semi-micro analysis only due to the presence of limited number of 'exchange' groups. Some of the important applications of ion-exchanger are:

(a) Conversion of a sodium salt of a carboxylic acid into the free acid e.g., if the cation exchange resin is represented as $Res.SO_3^-H^+$.

$$Res.SO_3^-H^+ + CH_3COO^-Na^+ \rightarrow Res.SO_3^-Na^+ + CH_3COOH$$

(b) Conversion of the salt of a weak base into the free base e.g. if the resin is represented as $(Res.NMe_3^+)OH^-$

$$(Res.NMe_3^+)OH^- + (C_6H_5NH_3)^+Cl^- \rightarrow (Res.NMe_3^+)Cl^- + (C_6H_5NH_3)^+OH^-$$

$$\downarrow$$

$$C_6H_5NH_2 + H_2O$$

(c) Separation of mixtures of acids and neutral compounds by the removal of acids.

(d) Separation of mixtures of bases and neutral compounds by the removal of bases.

2.13.4.2 Exercises in ion-exchange chromatography

Examples of each of the application discussed in the above section are:

(a) Conversion of sodium citrate into citric acid

This conversion cannot be carried out easily on a semi-micro scale by ordinary chemical methods. Dilute acids can be used to liberate an acid from its salts provided the organic acid is insoluble in water e.g. aromatic acids or it is volatile e.g. acetic acid. Citric acid is water-soluble and non-volatile and therefore ion-exchange method is used in this case.

Prepare a slurry of a strong acid cation resin by stirring it with 100-120 ml of 10% hydrochloric acid in a breaker and allowing it to settle. Decant the supernatant liquid. Put a small plug of glass wool at the base of a 50 ml burette. Now pour the slurry of resin to the burette to produce a column of 25-30 cm long. Open the tap of the burette and pass 100-120 ml of 10% hydrochloric acid through the resin at the rate of 5 ml per minute. Wash the resin with distilled water until the washings of the burette are neutral to methyl orange. Close the tap of the burette. Now column is ready for use.

Prepare a solution of 0.05 g of sodium citrate in 100 ml of water and pour it down the column at the rate of 5 ml per minute. Collect the runnings in a conical flask and wash the column with 100-150 ml of distilled water. Add these washings to the above runnings. Evaporate the combined runnings to obtain pure citric acid, having m.p.100°C.

For the regeneration of the column, pass excess of 10% hydrochloric acid and wash with water as described above. This column is now ready for reuse.

(b) Conversion of aniline hydrochloride into aniline

Prepare a column of a strong base anion resin of 25-30 cm long in a 50 ml burette as described in the above example. Wash the resin with distilled water and then pass 100-120 ml of N-sodium carbonate solution through column. Wash the column again with 200 ml distilled water. Now column is ready for the use.

Dissolve 0.05 g of aniline hydrochloride in 100 ml of distilled water and pass the solution down the column. Washings contain aniline in solution.

(c) Removal of acids from mixtures of acids and neutral substances

Removal of citric acid from the mixture of citric acid and glucose.

Prepare a column of a strong base anion resin as described in example (b). Prepare a solution of 0.025 g of citric acid and 0.25 g of glucose in 100 ml of distilled water. Pass the solution down the column at the rate of 5 ml per minute. Collect the runnings. Wash the colour with 100 ml distilled water. Add these washings to the above runnings and evaporate the water to obtain pure glucose. Citric acid is retained in the column and can be obtained by eluting the column with 100 ml of N-ammonium carbonate solution.

If required, regenerate the column [wash with water, followed by N-sodium carbonate solution and again with water].

(d) Removal of bases from mixtures of bases and neutral substances

Removal of benzylamine from the mixture of benzylamine and mannitol.

Prepare a column of a strong cation resin as described in example(a). Prepare a solution of 0.05 g of benzylamine and 0.01 g of mannitol in 100 ml of distilled water. Use the above procedure to obtain mannitol by the evaporation of the runnings.

2.13.5 High Performance Liquid Chromatography (HPLC)

Also called as High Pressure liquid chromatography. It is a widely used variation of column chromatography. The particles of the solid phase used in this chromatography are very finely divided in contrast to those used in column chromatography. Stationary phases specially prepared for adsorption, partition or ion-exchange chromatography are used in HPLC. The uniformity of grading of the very fine particles of the column packing gives the high resolution of the system.

In this chromatography, the analysis of mixtures of solid compounds may be carried out with a high rapidity and resolution as compared to that obtained with GLC. And, also, this technique can be used for relatively involatile or thermally unstable compounds, which cannot be used in GLC.

In this technique, stainless steel columns of 50 to 100 cm in length and 2-4 mm internal diameter are used. The finely divided nature of the solid phase and the above column dimensions, provide great resistance to the flow of the liquid and thus high pressures (1000-6000 p.s.i.) are required to force the eluant through the columns. The pressures are maintained by suitable pumps and can be preset at the required value.

2.14 ELECTROPHORESIS ON FILTER PAPER (FILTER PAPER ELECTROPHORESIS)

This technique is closely associated with the chromatography. Here separation is due to the difference in the electrical properties of the components of a mixture. Under the influence of an electric field, organic ions move towards an electrode (anode or cathode) with different rates and so results in the separation of mixtures. The pH of an electrolyte is very important and is maintained by a buffer solution.

The apparatus used in this technique is shown in the Figure 2.41. It consists of two terminals for the connection to power supply, shoulder piece for supporting the ends of the paper strip, paper strip holder to hold the paper and cotton wool plug, soaked in buffer solution, to make electrical connection between the compartments.

For the separation of mixtures by filter paper electrophoresis, first the compartments are filled by buffer solution (up to 10 mm depth). In each compartment, the level of the buffer solution must be same and if not so, tilt the tank and return to the original position. The filter paper connections, of the length of strip holders and width to dip well into the

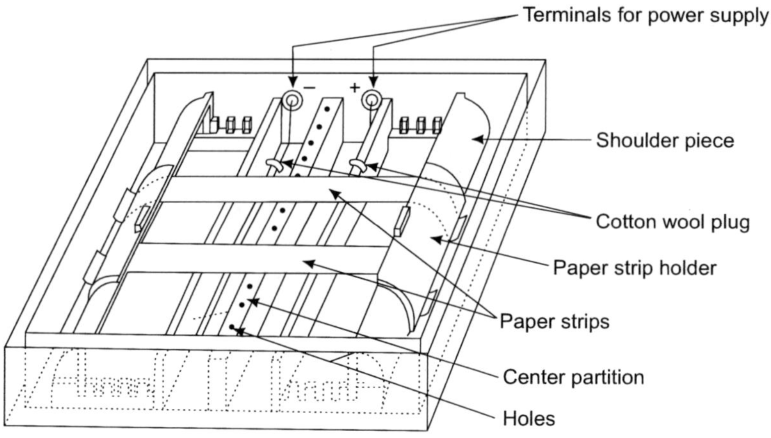

Fig. 2.41 Apparatus for electrophoresis on filter paper

buffer solution are prepared for strip holders. They are then lined with Whatman filter paper connections moistened with buffer by keeping the upper edge of the filter paper exactly at the edge of the holder. In the holes of centre partition, the support pins are inserted to support the edges of the filter paper strips. A base line is marked with a pencil on one-quarter length of each strip of 25 × 2.5 cm of Whatman No.1 filter paper. These paper strips are then moistened with buffer solution by dipping them in it followed by the removal of excess of solution by keeping them between the sheets of Whatman filter paper. Now the paper strips are inserted over the shoulder piece in such a way that the base line is towards the cathode and the paper strips are tightly secured by covering the ends with the paper strip holders.

The solution of the mixture to be separated is now applied on the base line across the width of the filter paper strip by means of a thin capillary tube and the ruler (to apply the solution in a straight line). The cover is placed in the position and the constant voltage source of 20-30 volts per inch of the effective strip length is now switched on. The current is switched off after 1-2 hours and cover is removed. Filter paper is removed with the help of forceps and the separation of components of the mixture on the strips can be observed.

2.14.1 Exercise on Filter Paper Electrophoresis

Separation of congo red, phenol red and bromophenol blue.

Prepare 0.1 percent aqueous solutions of each of the above indicators. Now prepare the mixture by taking equal volumes of the each of the above solutions. Also prepare buffer solutions of pH 4 & 10 by dissolving 1.5 g of ammonium acetate in 500 ml of 0.05 N acetic acid and 500 ml of 0.05 N aqueous ammonia respectively.

Follow the procedure as described above. Take out filter paper strips and dry them in an oven at 100°C for 5-10 minutes. For developing the spots, expose them to the fumes of concentrated hydrochloric acid followed by the fumes of aqueous ammonia solution (sp. gr. 0.88).

Similarly carry out the separation of the above indicators by using buffer solution of pH 10.

3

Isolation (or Extraction) of Products from Reaction Mixture/Natural Sources

3.1 ISOLATION OF PRODUCTS FROM REACTION MIXTURE

Synthetic organic chemistry involves various types of reactions. The most important step is the working up of the reaction mixture and isolation of the product or products. The products can be isolated by one of the following methods.

3.1.1 Isolation of a Solid by Filtration

The reaction mixture is poured on ice water and allowed to stand after stirring the mixture. If a solid product separates, it is collected by filtration using a fluted filter paper or by filtration using a Buchner funnel (or a filter disc) and suction. The separated product is dried and crystallised with a suitable solvent. The crystallised product is subjected to thin layer chromatography (TLC) to ascertain that it is a single pure product. If it is found that the product is a mixture of two or more compounds (which could not be separated by crystallisation), the mixture is purified by column chromatography.

Note: It is advisable to perform a TLC with the crude product so as to get an idea of the products present and their approximate amounts.

3.1.2 Isolation of a Liquid Product by Simple or Fractional Distillation

If the product obtained after pouring the reaction mixture in ice water is liquid, it is extracted with ether or dichloroethane. The liquid product is isolated by drying the solution (with a suitable drying agent), filtration (to remove the drying agent) and distilling the solvent on a stream bath/oil bath. The liquid product is obtained by distillation of the crude product. If the liquid product is a mixture (as indicated by TLC), it is obtained in pure state by redistillation or fractional distillation.

3.1.3 Isolation of Neutral/Acidic/Phenolic/Basic Impurities

The above two methods are recommended to be used for most of the classroom preparation. However, for working up other reactions, it is advisable to isolate the product by the following procedure. Generally, in any reaction we normally expect a product, which may be acidic, basic or neutral.

3.1.3.1 Isolation of a neutral product

The reaction mixture is poured into ice water and the mixture is extracted (using a separatory funnel) with ether or dichloromethane (this procedure

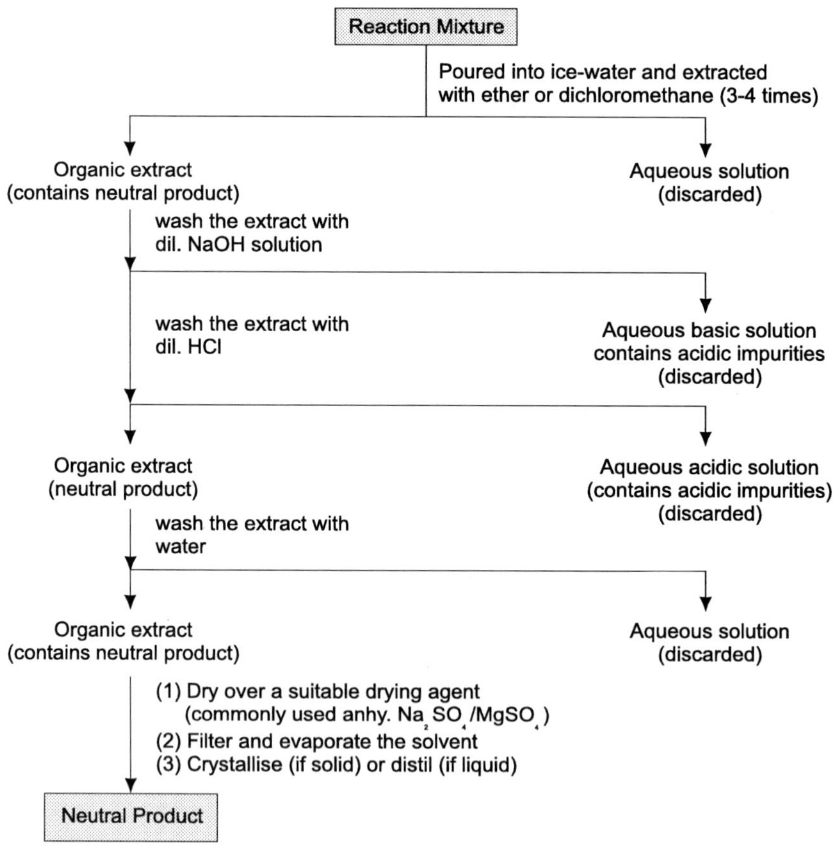

Fig. 3.1 Isolation and purification of neutral fraction

is used irrespective of whether the product is solid or liquid). The organic solvent is washed with dilute sodium hydroxide solution and then with dilute hydrochloric acid or dilute sulphuric acid. Finally the extract is washed with water and then dried over a suitable drying agent (anhydrous Na_2 SO_4/Mg SO_4) and finally the solvent is evaporated to leave the neutral compound as the residue. If it is solid, it can be crystallised. Alternatively, if the neutral compound is a liquid it is distilled* or fractionally distilled. The various steps are summarised in the diagram (Fig. 3.1).

3.1.3.2 Isolation of acidic product

The reaction mixture is poured into ice-water and the mixture is extracted with ether or dichloromethane (as in 3.1.3.1). The organic phase is extracted with sodium bi-carbonate solution. The sodium bicarbonate solution is extracted with ether or dichloromethane and the clear sodium bicarbonate solution is acidified with dilute hydrochloric acid. The acidified solution is extracted with ether or dichloromethane, washed with water and dried over a suitable dehydrating agent (anhydrous Na_2SO_4) and finally the solvent is evaporated to give acidic component. If it is solid, it is crystallized. Alternatively, if it is liquid, it is distilled or fractionally distilled. The various steps are summarized in the diagram (Fig. 3.2).

3.1.3.3 Isolation of phenolic product

The reaction mixture is poured into ice water and the mixture is extracted with ether or dichloromethane (as in 3.1.3.1). The organic phase is extracted with sodium bicarbonate solution. In case acidic component is absent (as seen in 3.1.3.2), the organic phase is extracted with dilute sodium hydroxide solution (it is presumed that the acidic component is absent). The sodium hydroxide solution is re-extracted with ether or dichloromethane and the clear alkaline solution is acidified with dilute hydrochloric acid. The extract is dried over a suitable dehydrating agent (anhydrous Na_2SO_4) and finally solvent evaporated off to give phenolic component. If it is solid, it is crystallised from a suitable solvent. Alternatively, if it is liquid, it is distilled or fractionally distilled. The various steps are summarised in the diagram (Figure 3.3).

*Depending upon boiling point, the distillation is performed at either atmospheric pressure or reduced pressure.

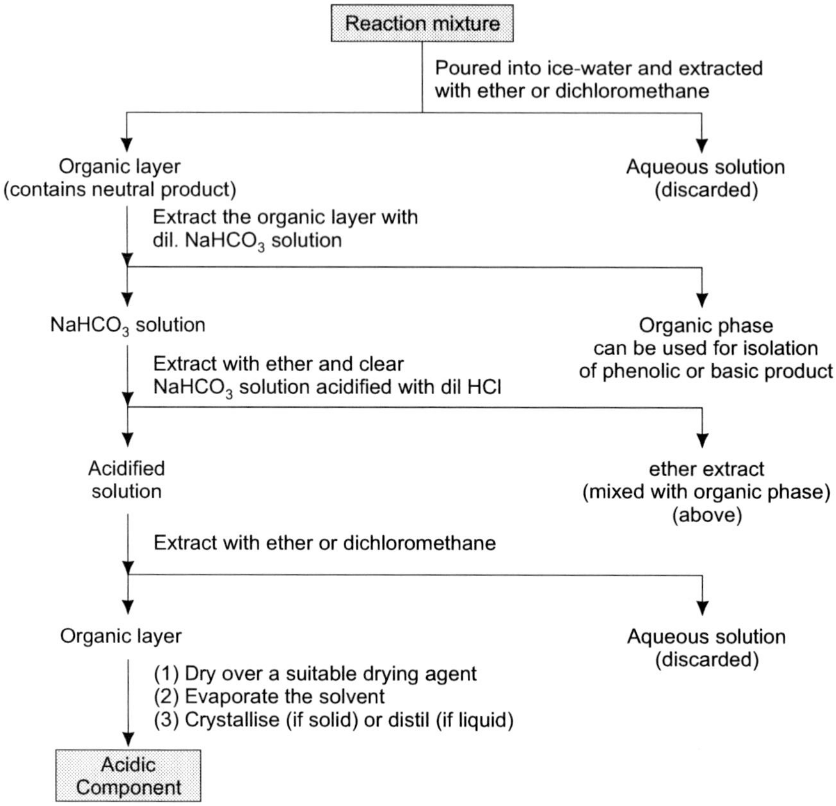

Fig. 3.2 Isolation and purification of acidic fraction

3.1.4 Isolation of Basic Product

The reaction mixture is poured into ice water and the mixture is extracted with ether or dichloromethane (as in (i) above). The organic phase is extracted with dil. hydrochloric acid. The lower acidic portion is re-extracted with ether and the clear acidic solution is basified with dilute sodium hydroxide solution. The basified solution is extracted with ether and the ether extract washed with water. It is dried over a suitable dehydrating agent (Anhy. Na_2SO_4), filtered and solvent removed. The residual product is crystallised if solid or distilled if liquid. The various steps are summarised in the diagram (Fig 3.4)

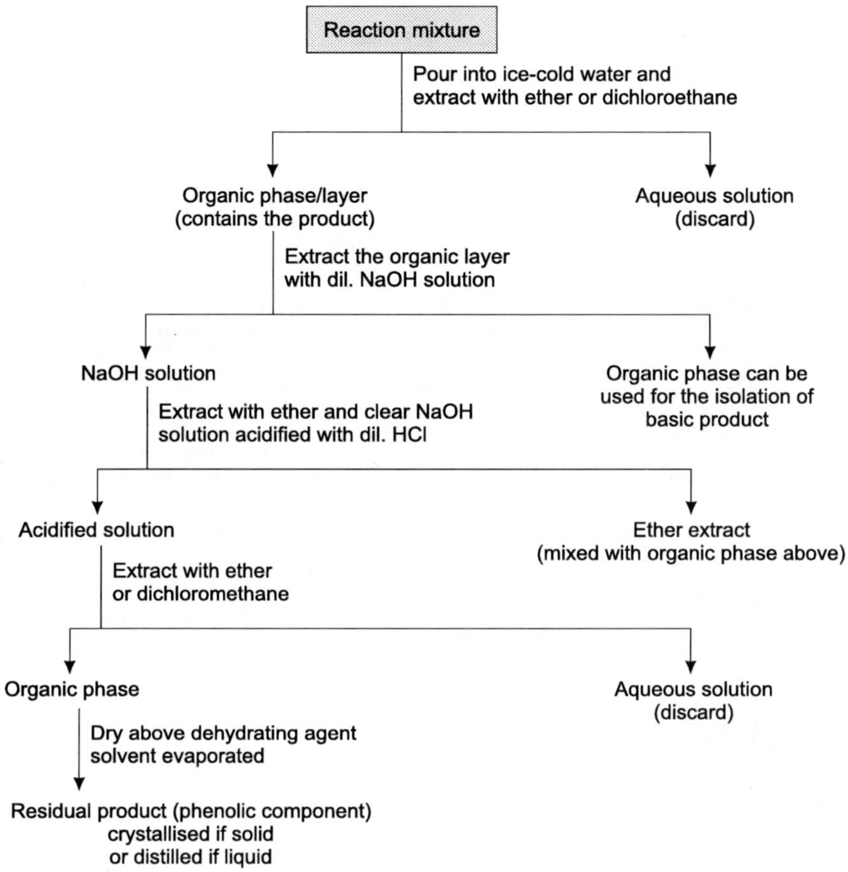

Fig. 3.3 Isolation and purification of phenolic component

The above method of separation of the reaction product into neutral, acidic, phenolic or basic fractions can also be used for separation of an organic mixture into various components.

In all the procedures used above, the organic compound is dissolved in ether or dichloromethane and dried using a suitable dehydrating agent before removal of the solvent. The dehydrating agent to be used must be insoluble in the organic solvent and should not react with the product. Following drying agents are generally used: anhydrous calcium chloride, calcium sulphate (Drierite), anhydrous magnesium sulphate, anhydrous sodium sulphate, anhydrous potassium carbonate and molecular sieves,

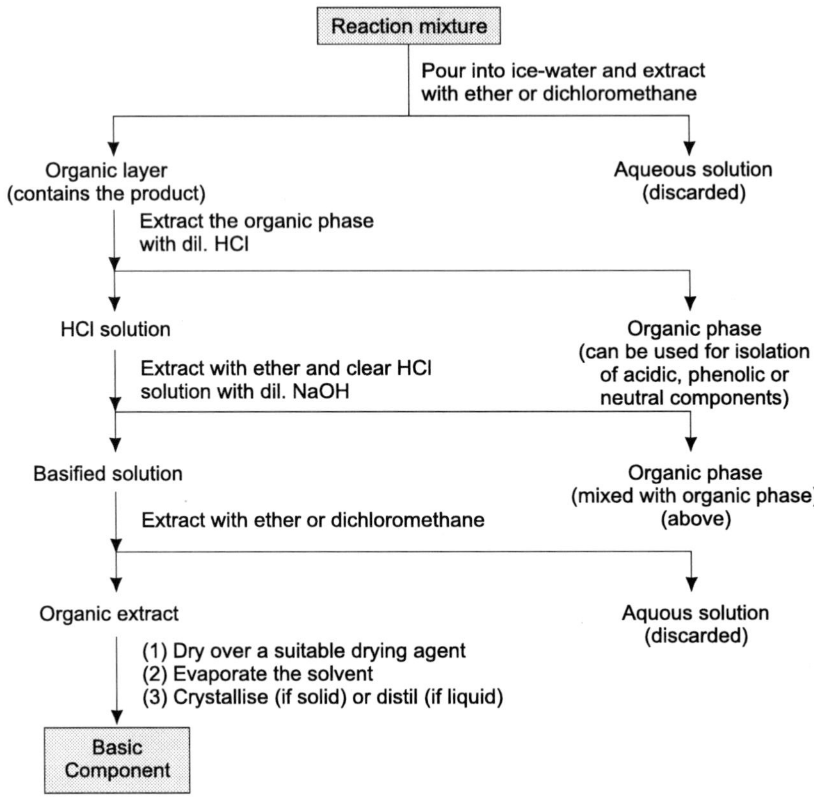

Fig. 3.4 Isolation and purification of basic fraction

4 Å. The choice of the dehydrating agent depends on the nature of the product in the solvent.

3.2 EXTRACTION OF PLANT MATERIALS

Plant materials (e.g. wood, bark, seeds, flowers, leaves, herbs, marine sources) are a source of large number of compounds, which may have varying degree of general use or biological activity. In view of this it is extremely important to investigate not only those plant materials which have not been investigated but also the plant materials which have been investigated earlier with a view to isolate additional or other minor components which may have different type of activities. Following are the methods, which have been used:

3.2.1 Extraction of Plant Material in Hot

The powdered and dried plant material is taken in a R.B. flask and the solvent is added . The flask is fitted with a reflux condenser and heated on a steam bath for 6-8 hr as shown in Fig. 3.5. After this the flask is cooled to room temperature and the solvent is decanted (filtered through cotton wool) and concentrated by distillation. The above process of extraction is repeated 4-6 times and the total extract is concentrated. The concentrate is then cooled to see if any solid material separates. Alternatively, it is taken up in ether and separated into acidic, phenolic, basic and neutral components as described earlier.

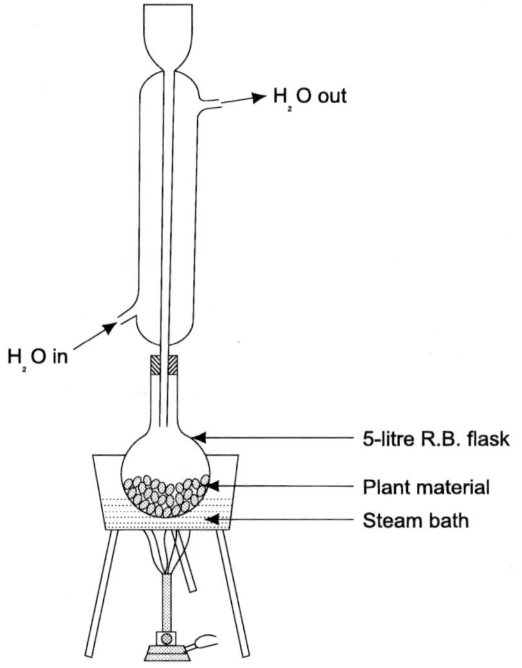

Fig. 3.5 Extraction of plant material

For extraction, the solvents are taken in increasing order of polarity. The order is petroleum ether (Hexane), ether, benzene, alcohol and acetone. Each extract is separately concentrated and the concentrate is worked up.

Alternatively a soxhlet extractor (Fig. 3.6) is used. The advantage of using a soxhlet apparatus is that while the plant material is extracted, simultaneously the extract is concentrated.

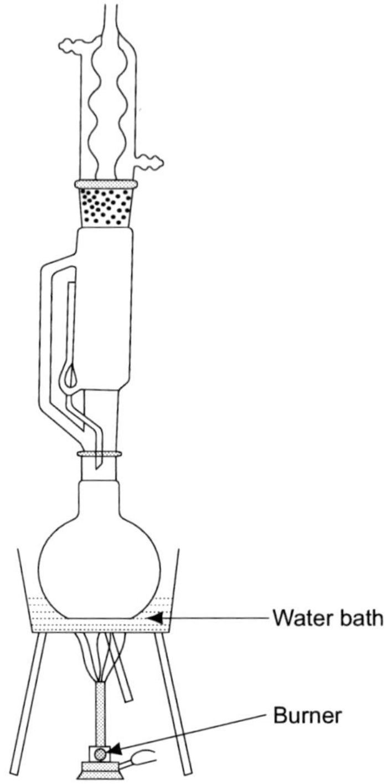

Fig. 3.6 Soxhlet extraction

The hot method of extraction of plant material is very convenient. However in certain cases, due to heat, some changes take place in the nature of the compound, e.g. isomerisation etc. In such cases the following method is used.

3.2.2 Extraction of Plant Material in Cold

The powdered and dried plant material is taken in an aspirator bottle and filled with the plant material, as shown in Fig. 3.7. The necessary solvent is added and the aspirator bottle allowed to stand overnight (10-12 hr) with occasional shaking (if possible). The solvent is removed by decantation (filtered through cotton wool) and concentrated in a rotary evaporator at

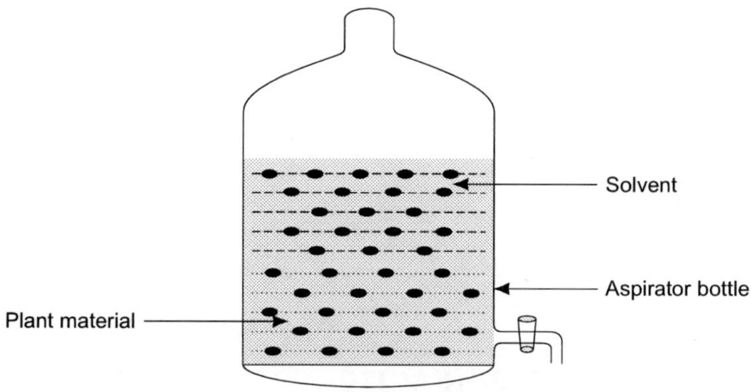

Fig. 3.7 Extraction of plant material in cold by using an aspirator bottle

as low temperature as possible and in vacuo (Fig. 1.56). The extraction is repeated 4-6 times and the total combined extract is concentrated in rotary evaporator. The concentrated extract is worked up as described earlier. Different solvents as described earlier are used for extraction and each concentrate is worked up separately.

3.2.3 Extraction of Plant Material in Cold with Acetone

The above two processes are quite tedious and consume lot of time in working up a large number of fractions. It has been found more advantageous to extract the plant material in cold with acetone (4-6 times). The total extract is concentrated in a rotary evaporator and taken in a thimble (made of hard filter), shown in Fig. 3.8 and put in the soxhlet apparatus. It is extracted successively with solvents in increasing order of polarity. The order normally followed is

Petroleum ether (removes oils and waxes)

Petroleum ether – Ethyl acetate (20%) mixture

Petroleum ether – Ethyl acetate (60%) mixture

Petroleum ether – Ethyl acetate (80%) mixture

Each of the concentrated extract is work up separately. Normally each extract is concentrated and screened as such for various activities. The extract, which shows maximum activity, is worked up to isolate the active principle.

Fig. 3.8 Thimble

3.3 LIQUID-LIQUID EXTRACTION

At times it may so happen that the concentrated extract obtained by the extraction of plant material in cold or hot does not give the products (neutral, acidic or phenolic) by the process described earlier. This may be because the products on treatment with alkali or carbonate may undergo change. In such cases it is most convenient to obtain pure products by liquid-liquid extraction procedure. In this method the crude oil material is fractionated using solvents for extraction in increasing order of polarity.

The process involves the use of a special device (see Fig 3.9) called liquid-liquid extractor. It is filled on to a R.B. flask (A) containing the oily extract. The solvent for extraction is taken in another R.B. flask (B) which is fixed on to the side arm of the extractor. A water condenser is fixed on the top of the liquid-liquid extractor. The side flask is heated on a steam bath. The solvent evaporates and goes through the side arm (C) and vapours get cooled in the water condenser and pass on to the bottom of the R.B. flask (A). The solvent extracts the organic compound. When the level of solvent reaches the mark, it goes to the flask (B). The process becomes continuous and the oil in the flask is thoroughly extracted. The extracted solvent is concentrated and the product is allowed to crystallise. In case it does not get crystallised, it is isolated by column chromatography. The solvent in the flask (B) is changed and the process of extraction is repeated.

Even during working up of a reaction, the product obtained may be soluble in water. In such cases also the liquid-liquid extraction process is extremely helpful.

If the oily product has to be extracted in hot, the flask (A) is heated on a water bath. The temperature of the water bath should be at least 5° lower than the boiling point of the solvent which is being used for extraction.

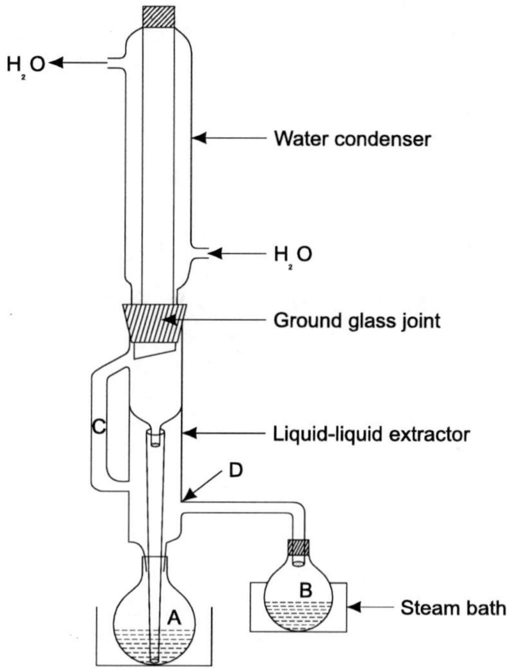

Fig. 3.9 Liquid-liquid extraction process

3.4 EXERCISES ON ISOLATIONS

3.4.1 Isolation of Acetylsalicylic Acid from Aspirin Tablets

Aspirin tablets consist of acetylsalicylic acid and a binder which binds the tablet.

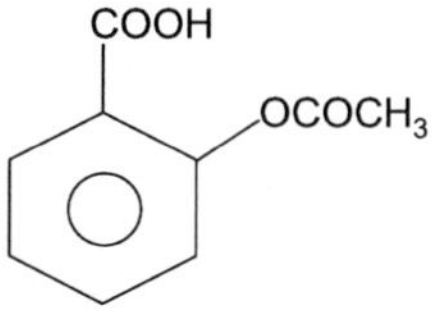

(Acetylsalicylic acid)

Place one 5-grain aspirin tablet (325 mg of acetylsalicylic acid) in a test tube and add 1 ml of 95% ethanol to it. Heat the test tube on steam bath to boil the mixture. Disintegrate the tablet with a glass rod. The acetylsalicylic acid will dissolve in the solution while the binder will remain as a white residue.

Filter the hot solution to remove binder (as shown in Fig. 2.2). Collect the filtrate in another test tube and add 5 ml of cold water to it. Stir the mixture, immediately acetylsalicylic acid will start to crystallize. Allow the mixture to stand for 10 minutes at room temperature and 5 minutes in an ice-bath for complete crystallization.

Filter the acetylsalicylic acid by vacuum filtration and wash the solid with a little cold water.

M. Pt. of acetylsalicylic acid = 135-36°C and Yield = 240 mg (75%)

3.4.2 Isolation of Ibuprofen from Ibuprofen Tablets

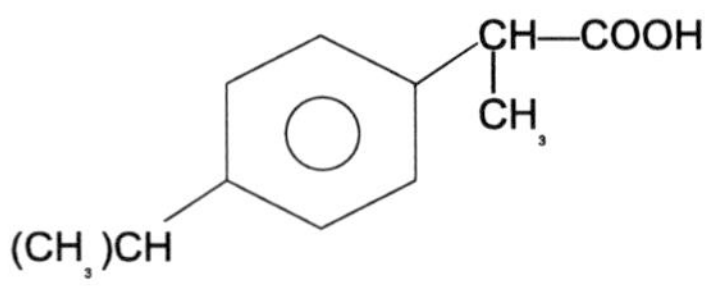

(Ibuprofen)

Peel the coating from 5 coated ibuprofen tablets (1 g of ibuprofen) with the help of metal spatula. If the coating doesn't peel off easily, crush the tablet which will disintegrate when the mixture is boiled. Place them in a 100 ml round-bottom flask and add 30 ml of ethyl ether. Reflux it on steam bath until the tablets disintegrate. Ibuprofen will dissolve in the solution while the binder will remain undissolved. Filter the mixture to remove undissolved binder. Collect the filtrate in another round-bottommed flask and distil off all the ether. Alternatively, the solvent can be removed using a rotary evaporator. Allow the residue to cool and scratch it to induce crystallization.

M. Pt. of Ibuprofen = 70-75°C and Yield = 0.8 g (80%).

3.4.3 Isolation of Cholesterol from Gallstones

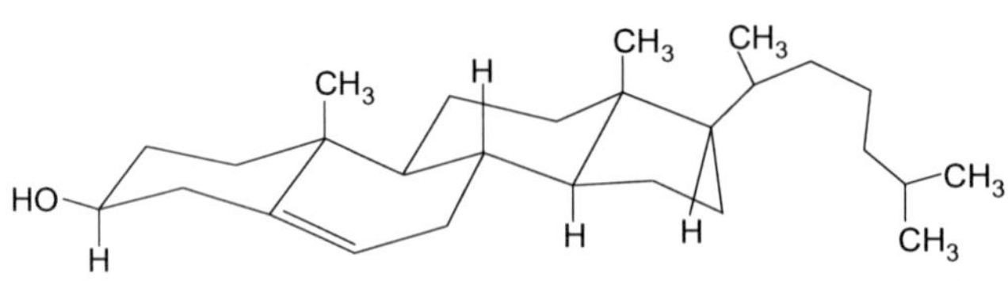

(Chloesterol)

Generally, human gallstones consist of about 75% cholesterol content.

Place 2 g of crushed gallstones in a 50 ml Erlenmeyer flask and add 15 ml of 2-butanone (or 15 ml of isopropyl alcohol or 10 ml of diethyl ether) in it. Heat the mixture on steam bath for about 5 minutes with occasional stirring. Filter the hot solution to remove the brown residue of bile pigments. Collect the filtrate in another 50 ml Erlenmeyer flask and dilute it with 10 ml of methanol. Add decolourising carbon to it and warm the mixture on steam bath for 1-2 minutes.

Filter the hot solution to remove carbon and collect the filtrate in another Erlenmeyer flask. Heat the filtrate on steam bath and add 1 ml of hot water into it. Swirl it to dissolve any precipitated solid. If all the precipitate doesn't dissolve, add a little methanol and heat with shaking to dissolve the precipitate. Allow the solution to cool, cholesterol will separate out as colourless plates.

Filter the crystals of cholesterol by vacuum filtration and wash them with a little of cold methanol. Recrystallize cholesterol from methanol.

M. Pt. of cholsterol = 149°C

3.4.4 Isolation of Caffeine from Tea

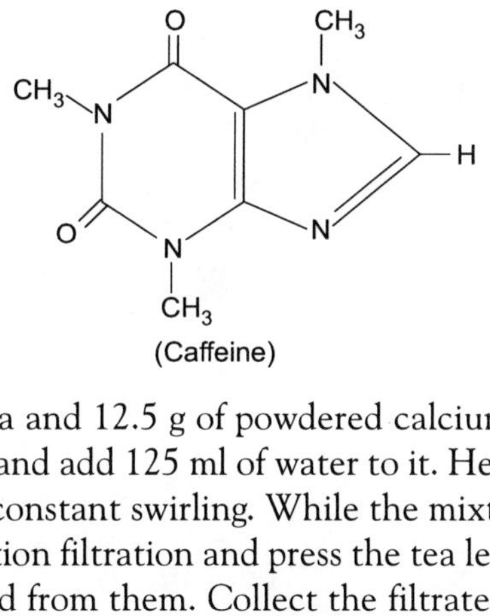

(Caffeine)

Place 12.5 g of tea and 12.5 g of powdered calcium carbonate in a 500 ml Erlenmeyer flask and add 125 ml of water to it. Heat the mixture for about 20 minutes with constant swirling. While the mixture is still hot, filter the solution with suction filtration and press the tea leaves with a large cork to take out the liquid from them. Collect the filtrate in a 250 ml Erlenmeyer flask and add 100 ml of dichloromethane. Stir the mixture containing two layers gently for about 10 minutes as vigorous shaking results in the formation of troublesome emulsions. Separate the two layers by using a separatory funnel.

Distil the chloroform layer to reduce its volume to about 10 ml and transfer the remaining solution to a weighed 25 ml Erlenmeyer flask (if the solution is not clear, filter it in the weighed flask). Evaporate the solution to dryness on the steam bath. Crystallize the caffeine from 95% ethanol [5 ml of ethanol per gram of caffeine]. Remove the greenish colour of the crystals by careful washing of the crystals with ethanol.

M. Pt. of caffeine = 235-36°C.

3.4.5 Isolation of Piperine from Black Pepper

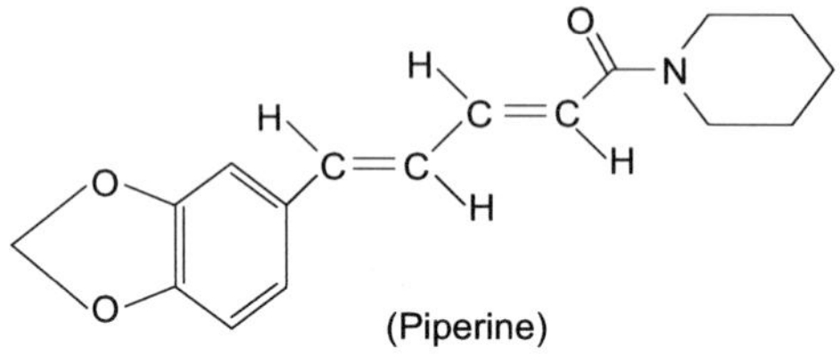

(Piperine)

Black pepper contains piperine (a very weakly basic substance) for about 10% of its weight.

Place 15 g of ground black pepper and 1 g of powdered calcium carbonate (to prevent the extraction of acidic components of pepper) in a 250 ml round-bottommed flask. Add 100 ml isopropyl alcohol in it and reflux the mixture for about 1 hr on steam bath.

Filter the hot mixture into a 125 ml Erlenmeyer flask and transfer it to another 250 ml round-bottommed flask. Distil the isopropyl alcohol to reduce the volume of the solution to about 10 ml. Transfer the residual solution to a 25 ml Erlenmeyer flask and allow the flask to cool for at least 24 hr for complete crystallization as crystallization occurs very slowly. An alternative for crystallization is, add 25 ml of water to the residual solution and allow the mixture to stand for at least 24 hr to complete the precipitation. Collect the solid by vacuum filteration. Recrystallize the piperine from either isopropyl alcohol or acetone.

Yield $\simeq$ 0.5 g

3.4.6 Isolation of Eugenol from Clove Oil

The characteristic odour of cloves is due to the Eugenol content.

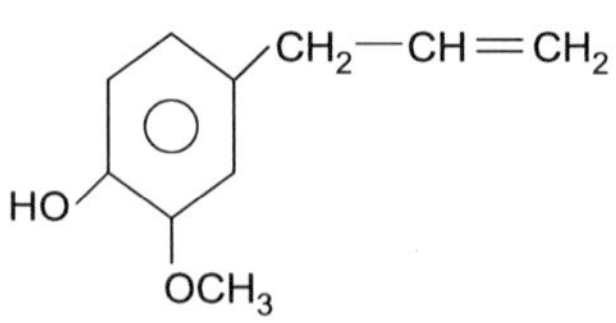

(Eugenol)

Dissolve 5 ml of clove oil in 50 ml of dichloroethane and extract the solution with three 50 ml portions of 5% aqueous potassium hydroxide solution by using a separatory funnel. Combine and acidify all the basic extracts with 5% aqueous hydrochloric acid (about 100 ml will be required). Extract eugenol from the above mixture with 50 ml of dichloromethane and separate the extract by using a separatory funnel. Dry the dichloromethane containing eugenol over anhydrous magnesium sulphate. Distil out the dichloromethane over a steam bath to get a residue of eugenol. B.Pt of Eugenol = 255°C

4

Criteria of Purity

Organic compounds can be derived either by natural sources or by preparation in the laboratory. By either of the way, they are contaminated with other compounds i.e. impurities. Before determining the properties of the compound, it is essential to obtain the compound in a state of purity. Purification of the organic compound can be carried out by various ways as discussed in Chapter 2.

The properties, physical as well as chemical, of a substance are determined by the molecular structure of the substance. Thus, the compounds having identical physical and chemical properties must be identical in their molecular structure and the substances having different properties must differ in the molecular structure. As a result, by comparing the physical and chemical properties of substances, we can tell whether they are the same.

When a sample is pure, its physical and chemical properties serve as the standards by which another sample of that substance can be identified. On the other hand, if a sample is impure, its purification is carried out by comparing its properties before and after the purification techniques.

The identity and purity of the substances can be established by certain physical properties i.e. M.Pt./B.Pt., mixed m.pt., TLC, superimposable I.R. Spectra.

4.1 CALIBRATION OF THERMOMETER

Thermometers should be calibrated before taking m.pts./b.pts of the given substance, as they are subject to errors due to the exposed stem or irregularities in the bore or inaccurate scale graduations.

Low cost long-scale thermometer (250-300 mm long) is usually calibrated for complete immersion of the mercury column in the vapour or liquid and thus, if such a thermometer is used, the entire mercury column is neither surrounded by the vapour nor completely immersed in the liquid and as a result a part of the mercury column is exposed to the cooler air of the laboratory and is not expanded as much as the bulk of the mercury and hence lower reading will be recorded. The error in thermometer due to the exposed stem can be corrected by adding a stem correction, which can be calculated by the formula.

Stem correction (in degrees) = $K\,N\,(t_2 - t_1)$

Where, K = Apparent expansion coefficient of mercury in glass and its value used in the above formula is 0.00016.

N = Length of the exposed stem

t_1 = Observed temperature

t_2 = Mean temperature of the exposed mercury column.

The error due to this exposed stem is very low upto 100°C, 3-5° upto 200°C & 6-10° at 250°C.

Due to various errors in thermometer it is essential to calibrate the thermometer. The calibration is carried out by taking m.pts. of pure solids or b.pts. of pure liquids. A calibration curve is then plotted on graph paper by taking temperature as abscissae and the corrections (to be added or subtracted) as ordinates. With the help of this curve, the thermometer correction at any temperature can be noted down.

Table 4.1 Reference Temperatures for Calibration of Thermometers

Melting Point (°C)		*Boiling Point (°C) at 1 atm pressure*	
Diphenylamine	54	Acetone	56
Naphthalene	80	Benzene	80
Oxalic acid (dihydrate)	101	Water	100
Benzoic acid	121	Chlorobenzene	132
Salicylic acid	158	Bromobenzene	156
Succinic acid	185	Benzaldehyde	179
3,5-Dinitrobenzoic acid	205	Acetophenone	202
		Methyl Salicylate	223
p-Chlorobenzoic acid	243	m-Bromoaniline	251
		I-Bromonaphthelene	281
Fumaric acid	286	Benzophenone	306

4.2 MELTING POINT DETERMINATION

The m.pt. of a solid is defined as the minimum temperature at which the solid and liquid are in equilibrium under one atmospheric pressure. The change in volume on the melting of a solid is very small in contrast to the change in volume on vapourization of a liquid. Thus, the m.pt. of a liquid is practically independent of any ordinary pressure change.

For a pure crystalline substance, the change from the solid to liquid state is sharp and also as the m.pt. determination is done accurately with even very small amount of substance, it is a valuable physical property often used for the identification and characterisation of the solids. As the addition of an impurity to a pure compound lowers down the m.pt. and increases the m.pt. range and therefore m.pt. is an important criterion of purity.

A sharp and definite m.pt. of an organic compound indicates its purity, with few exceptions, e.g. a eutectic mixture of two or more compounds may have a sharp m.pt. It can be differentiated from a pure organic compound by fractional crystallization from a suitable solvent(s). In practice, the exceptions are very small. Thus, a compound is said to be pure when it has a sharp and definite m.pt. which is unaffected by repeated fractional crystallisation.

4.2.1 Experimental Determination of Melting Point of a Solid

There are various ways by which the m.pt. of compound can be determined and the choice of the method largely depends on the amount of material available.

(a) Capillary melting points

A very minute amount of compound is needed in this method. Most commonly used method to determine the m.pt. is to introduce the substance into a capillary tube and then to heat it in a suitable bath of liquid and to note down the temperature at which melting occurs.

The apparatus employed for the determination of the melting point of a solid is shown in the Figure 4.1.

One end of the capillary tube is sealed by rotating it horizontally into the extreme edge of a small Bunsen flame for few seconds. The formation of a glass bead at the end of capillary tube should be avoided. A small amount (0.1 gm) of the dry compound is taken on a porcelain plate and finely powdered with a clean spatula, which is then inserted into the

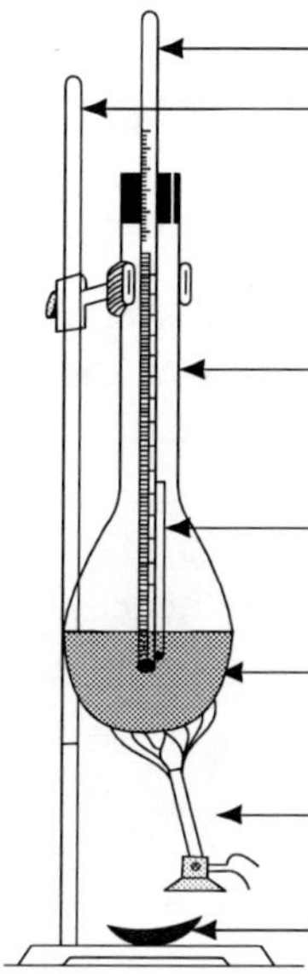

Fig. 4.1 Apparatus for melting point determination

capillary tube by pushing the open end of it into the powder and forcing the compound with the help of a spatula. The capillary should be filled compact by gentle tapping the close end on the hard surface (3-5 mm of the tube should be packed) by repeating the above procedure.

The filled capillary tube is now attached to the lower end of a thermometer, which is already wetted with the bath liquid. Following bath liquids can be used:

(1) Concentrated sulphuric acid is generally used as bath liquid and can be heated upto 250-280°C in a closed system while in an open container the temperature can be reached only to 200-220°C as above this temperature, excess of fumes are obtained. However, excessive fuming can be avoided to some extent by taking 70 parts by weight of conc. H_2SO_4 and 30 parts by weight of K_2SO_4 (or 55 parts by weight of conc. H_2SO_4 and 45 parts by weight of $KHSO_4$) and heating together in fuming cupboard till the whole solid gets dissolved. The addition of small amount of sodium or potassium nitrate can be used to decolourize the sulphuric acid and obtain its original colour.

(2) Paraffin: Can be heated to about 220°C as above this temperature it starts decomposing.
(3) Mineral Oil: The temperature can be reached upto 220°C.
(4) Silicone Oil: Has high flashpoints and so can be used at higher temperature.

The capillary gets attached on the wet thermometer by the surface tension of bath liquid by capillary attraction. Now the thermometer along with capillary tube is placed in the centre of the bath, by taking the precaution that sulphuric acid should not enter in the capillary tube. The Kjeldhal's flask is heated gently with the help of a burner. The rate of heating of bath should be uniform and slow near the m.pt. so that the thermometer and sample must be at the same temperature, otherwise the temperature of bath liquid will not be the same as the temperature of thermometer and sample. The temperature at which the sample melts is recorded as m.pt. of the sample. For a pure compound, the m.pt. range should be 0.5-1°C. If the approximate m.pt. is not known then first it should be determined by raising the temperature of bath quickly and then the accurate m.pt. is determined by low rate of heating near melting point. The two melting point determinations (fast & slow) will take less time than to determine the m.pt. of compound slowly over a wide range.

The second determination of m.pt. should be carried out by lowering the temperature of the liquid bath to about 30°-40°C and then replace the capillary tube by another fresh capillary tube filled with the compound whose m.pt. is determined and then the exact m.pt. is determined. The same capillary tube (when the solidification of the material takes place on cooling of the bath liquid) should not be used for second determination of m.pt. as it is possible that the compound may be partially decomposed or may be changed into another crystalline form having a different m.pt.

If the compound starts to decompose near its m.pt., the sample i.e. capillary tube should be placed in the bath only 5 to 10°C below the melting point so that the time in which sample is heated, gets minimised so avoids decomposition.

(b) Melting points from cooling curves

In this method, the solid whose m.pt. is to be determined must be available in about 1 g. So if the large amounts of the compound are available then this is the most accurate method. The apparatus used in this method is shown in the Figure in 4.2.

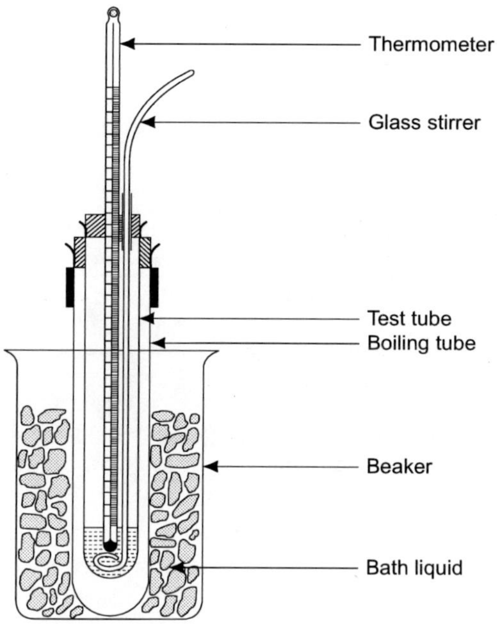

Fig. 4.2 Apparatus for melting point determination by cooling curve method

It consists of a test tube supported in a boiling tube by a cork, which prevents a rapid change of temperature, which is then inserted in a beaker having bath liquid. The test tube is having a thermometer and a glass stirrer. The compound whose m.pt. is to be determined is taken in this test tube. Sufficient amount of the compound is taken so that the bulb of the thermometer is completely dipped in its molten state.

In this method, the m.pt. is determined by first heating the sample till whole of it gets melted and then cool it slowly and record the temperature reading in equal fraction of time. On plotting the graph, temperature (ordinates) versus time (abscissae), at first, there is fall in temperature, as the liquid gives heat to the surrounding and then when crystallization begins, the temperature is constant until crystallization is completed, it is due to the fact that the heat evolved (heat of fusion) during this process will maintain the temperature at a constant value and then again fall in temperature as the solid loses heat to the surroundings. If the substance is pure, the temperature during the crystallisation process is horizontal and corresponds to the m.pt of the solid.

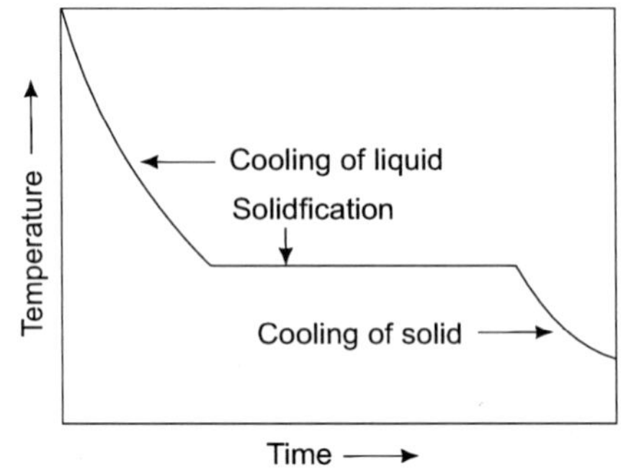

Fig. 4.3 Cooling curve for a pure compound

One more advantage of this method is that it can be used to determine the melting point of a substance, which is liquid at room temperature, i.e. its m.pt. is just above or below the room temperature. In this method, the substance whose m.pt. is to be recorded is first melted by using warm water bath and then placed in the beaker containing cold water or ice or a freezing mixture, depending on the m.pt. of the substance so that rate of cooling should be 0.5°C per minute. The temperature is now recorded after regular interval of time by stirring the compound and cooling bath. In the cooling curve, the horizontal portion will be the freezing point of the compound. For pure compounds, the melting point and freezing point are identical.

(c) Melting point determination by electrical heating

In this melting point apparatus, both the samples and thermometer are supported in electrically heated aluminium or copper block and the sample in the capillary can be observed through a magnifying glass. The block is heated electrically and the rate of heating can be regulated. The advantage of this method is that the temperature can be raised very slowly near the melting point since heat transfer is less than in liquid baths.

The features of this melting point apparatus are shown in Figure 4.4.

(d) Micro-hot-stage melting points

The advantage of this type of melting point apparatus is that m.pt of even a single crystal can be determined quickly, conveniently and with a better control of temperature. And also the nature and colour of crystals can be observed, during the entire temperature range.

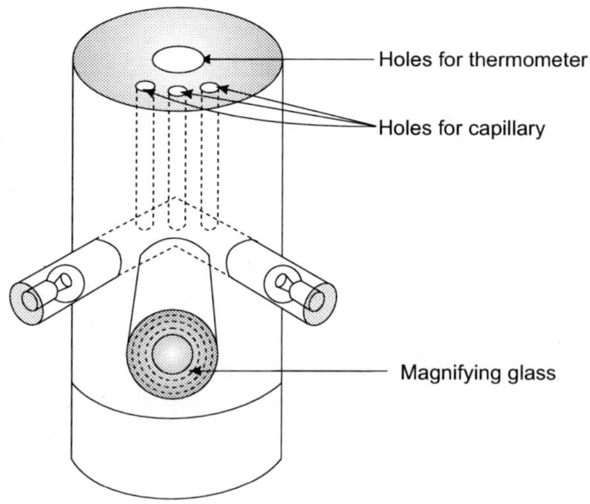

Fig. 4.4 Electrically heated melting point apparatus

The melting point is determined by heating the crystal of the solid between a pair of microscope cover glasses on an electrically heated metal block and observing the crystal through a magnifying glass.

However, the melting points recorded by this method are often higher than the capillary melting points as the thermometer is inside the block and the sample is on the surface and hence exposed to the cooler atmosphere.

4.2.2 Mixed Melting Point

In most of the cases, the impurity i.e. the presence of a 'foreign substance' lowers the melting point of a pure compound and it also melts over a wide range of temperature. Thus, this fact can be used to establish the identification of the organic compound. The method is called mixed m.pt. method, e.g. if an unknown compound is suspected to be benzoic acid (m.pt = 121°C) then it is mixed with equal amount of the pure sample of benzoic acid and the m.pt. of the mixture i.e. mixed m.pt. is determined. If the mixed m.pt. is again sharp and have the value, 121°C then the unknown compound is benzoic acid and if the mixed m.pt. is having a wide range and there is depression in the m.pt. value, then the compound is not the benzoic acid.

Thus, this method is a valuable method to identify the organic compounds. But it has some limitations also:

(a) Authentic sample may not be available for comparison.

(b) Since there are few exceptions to this method and so further proof of identity is essential.

Non-identity of two substances having the same m.pt. can be carried out by this method. Each individual compound melts at sharp and definite temperature and while a mixed m.pt. (m.pt. of the mixture of equal amount of two substances) is a wide range and is different from the m.pt. of the pure substances.

There are certain cases where the m.pt. of the mixtures are higher than the pure compounds, although such type of examples are rare. Examples of this class are: (a) If an additional compound of higher m.pt. is formed (b) If the two compounds are completely soluble in solid state and thus form solid solutions.

4.3 BOILING POINT DETERMINATION

The boiling point of a liquid is another valuable physical property to establish the identity and characterisation of the organic compound. Like m.pt., pure organic liquid boils at a fixed temperature but here impurities elevate the b.pt.

4.3.1 Experimental Determination of Boiling Point

It can be determined by two ways, depending upon the availability of the liquid. If sufficient amount is available, then the b.pt. of the liquid is determined by distillation. And if a small amount of the liquid is available then micro-boiling-point method (Siwoloboff's method) is used.

(a) Distillation method

It is the most accurate and hence preferable method to determine the boiling point, provided the substance is available in sufficient amounts. The temperature at which most of the liquid distills should be taken as the boiling point (Section 2.3). Since the boiling point is a function of pressure and thus, it should also be recorded.

(b) Siwoloboff's method

This method is used when a very small amount of substance (0.5 – 1.0 ml) is available. The apparatus used in this method is shown in the Figure 4.5.

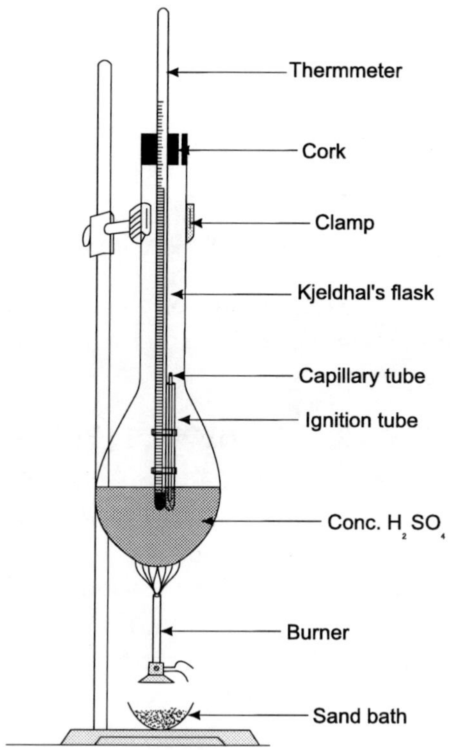

Fig. 4.5 Apparatus for boiling point determination

The apparatus consists of two tubes, one is an ignition tube (9-10 cm long and half a cm in diameter) and another is a capillary tube (8-10 cm long).

A small quantity of liquid, 3-4 drops, is added to the ignition tube, and the capillary tube sealed at one end is placed in the ignition tube with sealed end on the top. It is then attached to the thermometer with the help of a rubber band and the thermometer is dipped in the bath of a melting point apparatus, by taking precaution that rubber band should be quite above the bath liquid. Now the bath is heated, gradually by the burner until a very rapid i.e. continuous flow of air bubbles start coming from the open end of the capillary. The temperature when a continuous flow of bubbles just starts from the capillary tube is recorded and it is the boiling point of liquid. But to record this temperature is difficult and may have an error. Thus, to avoid this error, the more accurate result may be obtained by removing the burner and allow the bath to cool down slowly. Record the temperature at which the air bubble just ceases and the liquid starts to enter the capillary at the boiling point of the liquid.

4.3.2 Boiling Point Correction

As boiling point depends on the barometric pressure and so any deviation in the barometric pressure (760 Torr) will change the boiling point also. The observed boiling point will be low if the atmospheric pressure is low. If the pressure is not 760 Torr, the b.pt. observed could be corrected to the boiling point that would be expected at 760 Torr. This boiling point correction may be made by taking about 0.5°C for each 10 Torr deviation of atmospheric pressure from 760 Torr.

4.4 CHROMATOGRAPHIC TECHNIQUES (PAPER, TLC AND GAS CHROMATOGRAPHY)

Chromatographic techniques cannot only be used for the isolation, purification and detection of the organic compounds (Section 2.13) but can also be used to check the purity of the compounds. This technique is mainly used for:

(a) To check the purity of a compound obtained by a single simple fractional distillation (generally the compound obtained by this process is not of high purity).

(b) To check the purity of the starting materials produced commercially. As they may be contaminated with impurities of chemical similarity e.g. isomers, homologues, and so lead to the impurities in the final product also.

TLC and Paper Chromatography is a rapid and convenient method to check the purity of the solid compounds, while gas chromatography can be used for a liquid product or a volatile solid. Single spot in TLC and Paper chromatography and a single peak in gas chromatography indicate the purity of the compound.

4.5 I.R. SPECTRA

In I.R. spectroscopy, the usual wavelength ranges from 2.5 μ to 15 μ (4000-660 cm^{-1}). The region from 0.8-2.5 μ (12500-4000 cm^{-1}) is called near infrared and the region from 15-200 μ (660-50 cm^{-1}) is called far infared.

In organic analysis, main uses of I.R. spectroscopy are:

1. **Identification of a compound:** The simplest use of I.R. spectra is to find out whether the two compounds are identical or not. The spectra of two non-identical compounds differ particularly in the finger print region, 7-11 μ (1400-660 cm^{-1}). As this region is associated with many absorptions caused by bending vibrations as well as absorptions caused by the characteristic stretching vibrations, e.g. C–C, C–O and C–N, so it is the characteristic of the compound. Thus, this is the region which is usually checked for the identification.

   Hence, a perfect similarity in the I.R. spectra of two compounds i.e. completely superimposable spectra [provided the spectra of the two compounds are obtained under the same physical conditions, e.g. in solid state or solution, as the absorption bands shift in the different physical states] indicates that they are identical. The most convenient way of comparison of the two I.R. spectra is to plot both the I.R. spectra on the same paper with different ink pens [as shown in fig. 4.6]

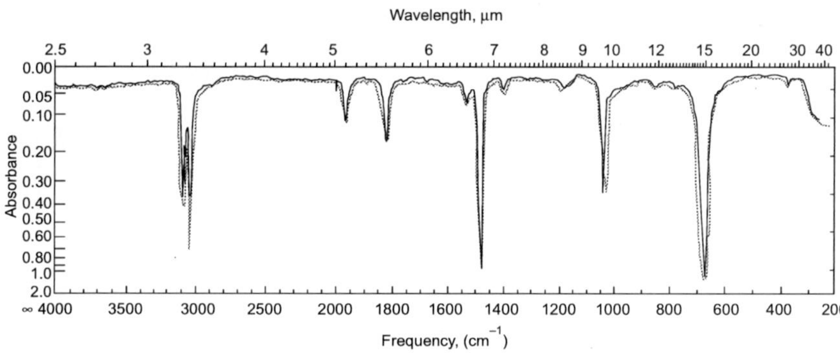

Fig. 4.6 I.R. spectra of two identical compounds with different ink pens.

2. **Determination of Purity:** For equal amount of the sample, the absorption peaks of the spectrum of an impure compound are broad while for a pure compound, they are intense [as shown in Fig. 4.7]. Also the spectrum of impure compound contains additional peaks.
3. **Identification of functional groups:** In I.R. spectroscopy, a particular group will have characteristic group of frequencies which

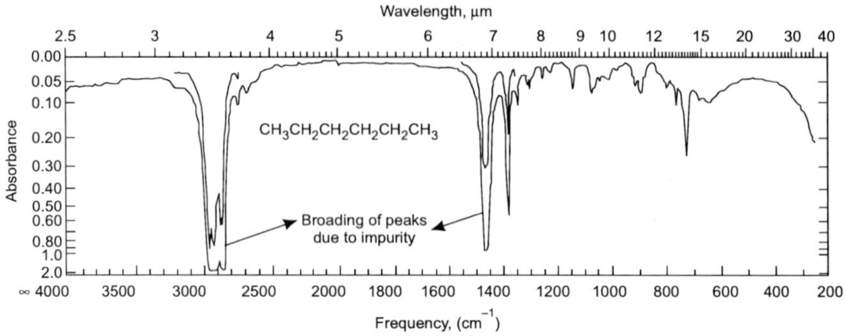

Fig. 4.7 I.R. spectra of impure compounds

are useful in the identification of that functional group e.g. a strong band at 2210 cm^{-1} indicates the presence of a cyanide group in the compound.

4. **Monitoring the purification:** I.R. spectra can also be used to monitor the purification by comparison of I.R. spectra of compound before and after recrystallisation or of different fractions obtained in a fractional distillation.

5

Solvents and Reagents

5.1 PURIFICATION OF SOLVENTS/REAGENTS

In any organic synthesis the solvents and reagents to be used must be pure and dry (unless the reaction is carried out in aqueous medium). The method consists of adding a suitable drying agent to the solvent or reagent, allowing to stand for some time, decanting and finally distilling. In certain cases the solvent is refluxed with the drying agent before distillation. The use of a particular drying agent depends on the nature of the solvent/ reagent to be dried. Following drying agents are generally used.

Table 5.1 Commonly used drying agents

S.No.	*Drying Agent*	*Suitable for drying*
1.	P_2O_5	Alkanes, Alkenes, Alkyl halides, Aromatic hydrocarbons, Aryl halides, Nitriles, ether
2.	KOH or NaOH	Amines
3.	BaO or CaO	Alcohols, Amines and ethers
4.	K_2CO_3	Aldehydes, Ketones, Amines, Nitriles
5.	$CaCl_2$	As in the case of P_2O_5, it cannot be used for drying nitriles
6.	$MgSO_4$	Alkyl halides, Aryl halides, ethers

Following are the methods/procedures to be adopted for purification of solvents/reagents in alphabetical order. The degree of purity to be achieved depends on the type of the reaction to be carried out.

5.1.1 Acetic Acid B.P. 118°, M.P. 16.6°

Commercial grade acetic acid is refluxed with potassium permanganate (2-5% by weight) for 4-5 hr to remove any acetaldehyde, and other oxidisable matter, which may be present. The acid is distilled and the fraction boiling at 117-118° is collected.

Any water present in acetic acid can be removed by cooling it in a refrigerator; the acid solidifies, excess water is removed by pouring. The process is repeated twice. The remaining acid is glacial acetic acid.

Last traces of water can be removed from acetic acid by treatment with triacetyl borate; it reacts with water to form acetic acid and boric acid (insoluble). The reagent is prepared by warming 1 part of boric acid with 5 parts (by weight) of acetic anhydride to 60°; on cooling triacetyl borate separates and is collected by filtration. (A. Pictet and A. Geleznoff, Ber., 1903, 31, 2219). Acetic acid is treated with 2-3 times of the amount of triacetyl borate estimated to be required for reaction with water present and anhydrous acetic acid is obtained by distillation.

5.1.2 Acetic Anhydride B.P. 140°

Absolutely pure and dry acetic anhydride is obtained by allowing it to stand over phosphorus pentoxide (50 gm/litre) for 2-3 hr. It is filtered and distilled in the absence of moisture.

5.1.3 Acetone B.P. 56.3°

It is refluxed (water bath) for 3-4 hr with solid potassium permanganate till the violet colour persists and then distilled. In case the distillate is coloured, it is redistilled using a small column (9") and finally dried over anhydrous ignited potassium carbonate. If necessary it can be redistilled before use, if the solvent has to be freed from traces of potassium carbonate.

5.1.4 Acetonitrile B.P. 81.6°

It is shaken with little phosphorus pentoxide (2% by weight) and distilled.

5.1.5 Anisole (Phenyl Methyl Ether) B.P. 153.7°

It is dried over anhydrous calcium chloride and then distilled over sodium.

5.1.6 Benzene B.P. 80.1°

Caution: Handle only in fume-hood. It is a long-term carcinogen and can cause leukemia. Whenever possible use less dangerous toluene or xylene instead.

It is purified by stirring mechanically crude benzene and concentrated sulphuric acid (80 ml per litre) for 30 min. The acid is separated (separatory funnel) and the acid treatment is repeated twice with fresh portions of acid. The Benzene layer is separated and distilled. Finally it is dried over sodium wire and distilled over phosphorus pentoxide. The method gives pure dry thiophene-free benzene.

5.1.7 Butan-1-ol B.P. 117.6°

It is purified by refluxing over calcium oxide and distilling. It is also purified by treatment with sodium (ca. 3 g per litre) and distillation. Alternatively it can also be purified by refluxing for 2 hr with dibutyl phthalate followed by distillation.

5.1.8 Butan-2-ol B.P. 99.9°

It is purified by distillation.

5.1.9 Tert-Butanol (2-methylpropan-2-ol) B.P. 82.4°, M.P. 25.8°

It is refluxed over calcium oxide for 2-3 hr and then distilled.

5.1.10 Iso-Butanol (2-methylpropan-1-ol) B.P. 107.6°

It is refluxed over calcium oxide (3-4 hr) and then distilled. It can be further dried with magnesium (see under ethyl alcohol, 5.1.26).

5.1.11 Carbon Disulphide B.P. 46.2°

It is distilled over phosphorus pentoxide (8-10 g per litre) and then distilled.

5.1.12 Carbon Tetrachloride B.P. 76.8°

It is refluxed over phosphorus pentoxide (5 gm per litre) for 30 min and then distilled.

5.1.13 Chlorobenzene B.P. 131.6°

It is shaken (3-4 times) with concentrated sulphuric acid, washed with water until neutral and distilled. The distilled product is dried over phosphorus pentoxide (5 g per litre) and redistilled. If required free of traces of acid, pass through an basic alumina column (activity I).

5.1.14 Chloroform B.P. 61.1°

Caution: In presence of light it reacts with oxygen to form phosgene ($COC1_2$) etc. Commercial chloroform is therefore stabilized by addition of ethanol (1%). Chloroform may react explosively with strong bases or alkali metals.

It is purified by shaking with concentrated sulphuric acid, washing with water, drying over calcium chloride and distilling. Finally it is distilled over phosphorus pentoxide (5 g/litre).

5.1.15 Cyclohexane B.P. 80.7°

It is shaken with concentrated sulphuric acid (5% by volume); the process is repeated 2-3 times, washed with water until neutral and then distilled. The distilled product is redistilled over phosphorus pentoxide.

5.1.16 Decalin (Decahydronapthalene) B.P. 191.7°

It is refluxed with sodium wire (ca. 1%) for 3-4 hr in an inert atmosphere (nitrogen) and then distilled. Finally it is passed through basic alumina column (activity I).

Caution: It forms peroxides in light and it should be stored under nitrogen.

5.1.17 1,2-Dichloroethane (Ethylene Chloride) B.P. 83.4°

It is distilled over phosphorus pentoxide and then filtered through basic alumina (activity I).

5.1.18 Dichloromethane (Methylene Chloride) B.P. 39.7°

It is refluxed for 30 min over phosphorus pentoxide (5 g per litre) and then distilled. Finally, traces of acid can be removed by filtration through basic alumina (activity I).

Caution: It reacts violently with alkali metals or strong bases.

5.1.19 Diethyl Ether (Ether) B.P. 34.5°

It is first dried over anhydrous calcium chloride and then over sodium wire and the clear decanted ether is distilled over a little phosphorus pentoxide in inert atmosphere. Ether should be stored in the absence of light.

Caution: In presence of light, ether combines with oxygen to form peroxides. So unpurified ether should never be distilled to dryness (otherwise the peroxides may explode violently). The presence of peroxides can be detected as follows: Shake 2-3 ml of ether with a solution of potassium iodide (200 mg) in hydrochloric acid (1N, 10 ml) – a brown colour (liberated iodine) indicates the presence of peroxide.

5.1.20 Diethylene Glycol Dimethyl Ether (Diglyme) $CH_3OCH_2CH_2CH_2OCH_3$ -B.P. 159° (decomp.)

Initial purification is done by distilling under reduced pressure (b.p. 62-63°/15 mm). Subsequently it is treated with sodium wire added at intervals until no further reaction is detected. It is decanted in inert atmosphere, allowed to stand for 3-4 hr over calcium hydride and then fractionally distilled under reduced pressure.

5.1.21 Dimethoxyethane (Monoglyme, Ethylene Glycol Dimethyl Ether) $CH_3OCH_2CH_2OCH_3$, B.P. 93°

It is treated with sodium wire added at intervals till no further reaction is detected. Finally the decanted solvent is fractionally distilled in inert atmosphere (nitrogen).

5.1.22 N.N-Dimethylformamide (DMF, Formdimethylamide) B.P. 153°

Caution: It decomposes slowly at 150°; the decomposition is base catalysed and proceeds at room temperature in presence of sodium hydroxide.

It is purified by mixing with benzene (30 g/250 g) (or cyclohexane, 30 g/250 g) and water (12 ml). The mixture is distilled at 140° to remove impurities together with benzene and water. Finally it is distilled under reduced pressure and stored in dark over type 4 Å molecular sieve.

5.1.23 Dimethyl Sulphoxide (DMSO) B.P. 189°, M.P. 18.5°

It is dried over calcium sulphate and then distilled under reduced pressure (75°/12 mm).

5.1.24 Dioxan (1,4-Dioxan, p-Dioxan) B.P. 101.3°, M.P. 11.8°

Caution: It forms peroxides. For detection see diethyl ether (5.1.19).

It is purified by drying over potassium hydroxide, decanting and distilling over sodium and stored in inert atmosphere.

5.1.25 Diisopropyl Ether, $(CH_3)_2$ $CHOCH(CH_3)_2$ B.P. 68.3°

Caution: It readily forms peroxides with oxygen.

It is purified as in diethyl ether (5.1.19).

5.1.26 Ethyl Alcohol (Ethanol) B.P. 78.2°

Ethyl alcohol of approx. 99.5% purity is prepared as follows: Ethyl alcohol (ca 95%) is refluxed for 3-4 hr with fresh quick lime (broken into small lumps). It is then allowed to stand for 2-3 days when the lumps disintegrate into powder. It is refluxed for 1 hr and distilled (in the absence of moisture).

Absolute alcohol is obtained as follows:

A mixture of magnesium turnings (5 g), alcohol (99.5%, 60 ml) is refluxed in a large flask until a vigorous reaction ensures and nearly all magnesium has been converted into the ethoxide. If the reaction is slow in the beginning, few crystals of iodine are added. Alcohol (99.5%, 900 ml) is added, mixture is refluxed for 1 hr (anhydrous conditions) and absolute alcohol is distilled.

5.1.27 Ethyl Acetate B.P. 77°

It is washed with sodium carbonate solution (5%) (to remove acetic acid), then with saturated calcium chloride solution, dried (anhydrous potassium carbonate) and distilled. Phosphorus pentoxide can also be used for final drying – the ethyl acetate is filtered and distilled.

5.1.28 Ethylene Glycol (Ethane-1,2-Diol) B.P. 197.3°

It is dried over anhydrous sodium sulphate and distilled under reduced pressure.

5.1.29 Ethylene Glycol Monoethyl Ether (Ethyl Cellosolve, 2-Ethoxy Ethanol) $CH_3\ CH_2\ OCH_2\ CH_2\ OH$, B.P. 135.6°

It is purified by distillation.

5.1.30 Methylene Glycol Monomethyl Ether (Methyl Cellosolve, 2-Methoxyethanol) $CH_3\ OCH_2\ CH_2\ OH$, B.P. 124.6°

It is purified by distillation.

5.1.31 Ethyl Formate B.P. 54.1°

It is dried over anhydrous magnesium sulphate, decanted and distilled over phosphorus pentoxide (10 g/litre).

5.1.32 Formamide B.P. 210.5° (decomp)

It is initially dried over anhydrous sodium sulphate and then distilled under reduced pressure (104°/10 mm).

5.1.33 GLYCEROL (GLYCERINE), 1,2,3-TRIHYDROXYPROPANE B.P. 290° (decompose)

Anhydrous glycerol is prepared by heating commercial glycerol in a porcelain-evaporating dish carefully over a wire gauze (in a fume cupboard). It is stirred steadily with a thermometer until the temperature rises to 180°. It is cooled to 100°, poured into a pyrex beaker and the beaker is kept in a large desiccator containing concentrated sulphuric acid. It is an extremely hygroscopic substance. The dried glycerol can be used in Skraup's quinoline synthesis.

It can, however, be further purified by high vacuum distillation.

5.1.34 Hexamethylphosphotriamide (HMPA), (Hexamethylphosphoramide) B.P. 233°, M.P. 7.2°

Caution: It is carcinogenic.

It is purified by drying over calcium oxide (24 hr room temperature) followed by high vacuum distillation.

5.1.35 n-Hexane B.P. 68.7°

Commercial hexane is a mixture of hydrocarbons containing n-hexane together with other alkanes, alkenes and less volatile materials.

It is purified by repeatedly shaking with concentrated sulphuric acid (5% by volume) (to remove alkenes), washing with water (until neutral) and fractional distillation over phosphorus pentoxide.

5.1.36 Methyl Alcohol (Methanol) B.P. 64.7°

Caution: It is poisonous.

It is purified initially by distilling over sodium (3-4 g per litre). Subsequently it is purified as follows. Place dry magnesium turnings (5 g/litre) in a flask and add 50 ml sodium distilled methanol. Wait until reaction starts (cloudiness; a trace of iodine may accelerate this step), then add the methanol requiring purification. The mixture is refluxed for 3-4 hr and then distilled.

5.1.37 Methyl Acetate B.P. 56.3°

It is purified as in the case of ethyl acetate (5.1.27).

5.1.38 Methyl Isobutyl Ketone (4-Methylpentan-2 one) B.P. 116.5°

It is dried over calcium sulphate (20 g/100 ml) and then fractionally distilled.

5.1.39 Nitrobenzene B.P. 210.8°, M.P. 5.7°

It is dried over phosphorus pentoxide (50 g/litre), decanted and then distilled in vacuum (b.p. 85°/7 mm).

5.1.40 n-Pentane B.P. 36°

It is purified as in hexane (5.1.35).

5.1.41 Petroleum Ether (Light Petroleum, Ligroin)

It is a mixture of hydrocarbons boiling over a particular range (e.g. 40-60°, 60-80°, 80-100°; it is not an ether). It is used as a substitute for pentane or hexane. It is purified as n-hexane (5.1.35).

5.1.42 Iso-Propanol (Iso-Propylalcohol, Propan-2-ol) B.P. 82.2°

It is dried over calcium sulphate (200 g/litre), decanted, treated with sodium wire (8 g/litre) and heated slowly to boiling. When alkoxide formation is complete, isopropyl benzoate (35 ml/litre) is added and the mixture is refluxed for 3 hr and distilled. This method provides isopropanol suitable for Meerwein-Ponndorf reductions. For other procedures see under ethanol (5.1.26).

5.1.43 Pyridine B.P. 115.2°

It is first dried over solid potassium hydroxide (2-3 days) and then refluxed (4 hr) and distilled.

5.1.44 Quinoline B.P. 238°

It is first dried over solid potassium hydroxide pellet (24 hr) and then distilled.

5.1.45 Sulpholane B.P. 287.3° (decomp.), M.P. 28.4°

It is dried over potassium hydroxide (50 g/litre) for 24 hr, decanted and distilled in vacuo. The pure material is solid at room temperature.

5.1.46 Tetrachloroethane, $CHC1_2$ $CHC1_2$, B.P. 146.3°

It is purified by warming on steam bath with concentrated sulphuric acid (80 ml/litre) and either shaken or stirred mechanically. The tetrachloroethane layer is separated and the process is repeated 2-3 times till the acid does not become coloured. The solvent is then separated, washed with water, dried (calcium chloride) and distilled.

5.1.47 Tetrahydrofuran B.P. 66°

It is purified by treatment with solid potassium hydroxide (24 hr) followed by distillation. It is finally distilled over lithium aluminium hydride.

A more recent method of purification involves refluxing the purified tetrahydrofuran with benzophenone and sodium or potassium under nitrogen followed by distillation [*Techniques of Chemistry*, A. Weissberger ed., Vol. 2, 3rd Edition, Wiley, 1970].

Caution: Peroxide formation is very rapid and so tetrahydrofuran should always be freshly distilled before use.

5.1.48 Tetralin (Tetrahydronaphthalene) B.P. 207.5°

It is refluxed for 1 hr over sodium (1%), distilled and finally passed through basic alumina (Grade I) and is stored under nitrogen. It forms peroxides readily.

5.1.49 Toluene (Methyl Benzene) B.P. 110.6°

It is purified as in the case of benzene (5.1.6).

5.1.50 Xylene

Commercial xylene is a mixture of isomeric xylenes b.p. ca 140° and purified by refluxing for 1 hr over sodium followed by distillation.

5.2 DRYING OF SOLVENTS

5.2.1 Use of Molecular Sieve

The term 'molecular sieve' applies to a group of dehydrated synthetic sodium and calcium aluminium silicate adsorbents and is also referred to as Zeolites. These have a crystal lattice structure having uniformly sized pores, which are able to accept molecules smaller than a limiting dimension. The larger molecules do not diffuse into the lattice structure. The selectivity of the action of molecular sieve is based on its molecular shape and dimension. This accounts for the sieving action and is particularly valuable in the removal of water from the solvents. The pore size of the molecular sieve is determined by the nature of the manufacturing process. A number of types are available in the market.

Drying of organic solvents is most efficient if pre-dried, distilled, peroxide-free solvent is percolated through a column of molecular sieve. A 250 g molecular sieve column (25 mm diameter – 600 mm high) can dry 10 litres of the following solvents to a purity better than 0.002% residual water: Benzene, chlorobenzene, cyclohexane, chloroform, carbon tetrachloride, diethyl ether, diisopropyl ether, decalin, dichloromethane, dimethyl sulphoxide, dioxane, ethyl acetate, tetrahydrofuran, toluene and pyridine (all these with 4 Å sieve) and acetone and acetonitrile (3 Å sieve). It is advisable to store the purified solvents in a stopper bottle containing molecular sieve (10-15 g/litre).

Molecular sieve quickly takes up atmospheric moisture and so should be handled carefully. It is always advantageous to use freshly activated molecular sieve.

Molecular sieve has the added advantage of capable of regeneration with no loss of activity. It should always be reactivated before use by heating in an oven at 300-350° either in vacuo or in a stream of nitrogen. Alternatively, the used molecular sieve is mixed with excess of water, filtered and dried well. It is then dried for 4-5 hr at 150° and then under high vacuum at 200° (silicon oil bath).

Caution: Peroxides cannot be removed by molecular sieve. However, these can be removed by chromatography over alumina (A.A. Morton and J.G. Mark, Ind. Eng. Chem., Anal. Ed., 1934, 6, 151).

5.3 SPECTROSCOPIC GRADE SOLVENTS

For making/purifying solvents for spectroscopy, the solvents, which have been purified, as described earlier (Sec 5.1) should be used. Following is given the detailed procedure, which is to be followed for making solvents for spectroscopy.

5.3.1 Acetonitrile B.P. 81.6°

The purified acetonitrile (Sec 5.1.4) is distilled using an efficient column (150 cm). The constant boiling fraction is refluxed for 1 hr with potassium permanganate (20 g per litre) and anhydrous potassium carbonate (20 g per litre) and then distilled rapidly. The distillate (which is basic and smells ammonia) is added dropwise to concentrated sulphuric acid until the acid end point. The acetonitrile is decanted from any inorganic impurities and distilled in an inert atmosphere (See Fig. 1.35) using an efficient column and discarding the forerun (5-10%).

5.3.2 Carbon Tetrachloride B.P. 76.8°

The carbon tetrachloride which has been purified (as described in Sec 5.1.12) is filtered through basic alumina (activity J) and then redistilled using a short column.

5.3.3 Chloroform B.P. 61.1°

The purified chloroform (Sec 5.1.14) is filtered through a small column (4-5") of basic alumina (activity I). Approximately 50 g of alumina will give ca 70 cm^3 chloroform containing less than 0.005% ethanol and traces of water or acid. The first 25 ml should be collected and passed through the column again. The stabiliser free chloroform is kept in a refrigerator in a dark bottle. It can be used for spectroscopic work up to two weeks.

5.3.4 Cyclohexane B.P. 80.7°

The cyclohexane which has been purified (as described in Sec 5.1.15) is filtered through basic alumina column (activity I).

5.3.5 n-Hexane B.P. 68.7°

The purified n-hexane (Sec 5.1.35) is distilled over sodium hydride and filtered through a column at basic alumina (activity I).

5.3.6 Methyl Alcohol (Methanol) B.P. 64.7°

The purified methanol (Sec 5.1.36) is refluxed for 4-5 hr with furfural (40 ml/litre) and sodium hydroxide solution (10%, 60 ml/litre). A resin is formed carrying down all the carbonyl compounds present. The remaining alcohol is decanted and then fractionated and separated from water and furfural. The distilled methanol is distilled over sodium (3-4 g per litre). Subsequently it is purified as follows:

To dry magnesium turnings (5 g/litre) in a flask is added 50 ml sodium distilled methanol. Within a few minutes the reaction starts (cloudiness starts appearing). A trace of iodine may accelerate this step. The remaining methanol is added, mixture is refluxed for 3-4 hr and finally distilled.

5.3.7 Methylene Chloride (Dichloromethane) B.P. 39.7°

The purified methylene chloride(Sec 5.1.18) is filtered through a small column at basic alumina (activity I) and then redistilled.

5.3.8 Water

Deionised water (obtained by passing through cation exchange and anion exchange resin respectively) is refluxed with potassium permanganate for 6-8 hr and then distilled. The distilled water obtained above is again refluxed for 6-8 hr with more of potassium permanganate and distilled again. Continuous distillation apparatus are available for this purpose.

5.4 SOLVENTS FOR HPLC

Solvents used for HPLC are of the same type as the solvents for spectroscopy. Before use, these solvents are treated as given below:

(i) Before use, the solvents are filtered through Nylon membrane and filters. These filters are generally available in 147 mm diameters and are fitted in special filtration arrangement as shown in the Fig. 5.1. These filters remove impurities of particle size 0.45 microns.

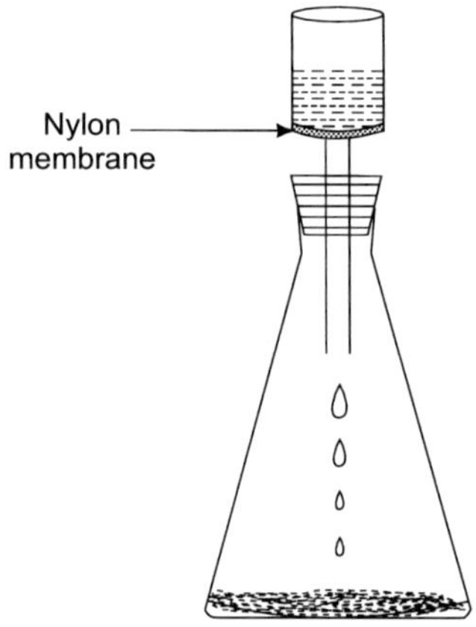

Fig. 5.1 Apparatus for filtration through nylon membrane

(ii) The sample for HPLC is dissolved in the required solvent, which is filtered through nylon membrane filter of 0.13 mm diameter. This filter is fitted in the syringe itself and removes impurities up to 0.2 micron particle size. One such arrangement is shown in the Fig. 5.2.

5.5 DEUTERATED SOLVENTS

Deuterated solvents are used for taking NMR spectra of compounds. The

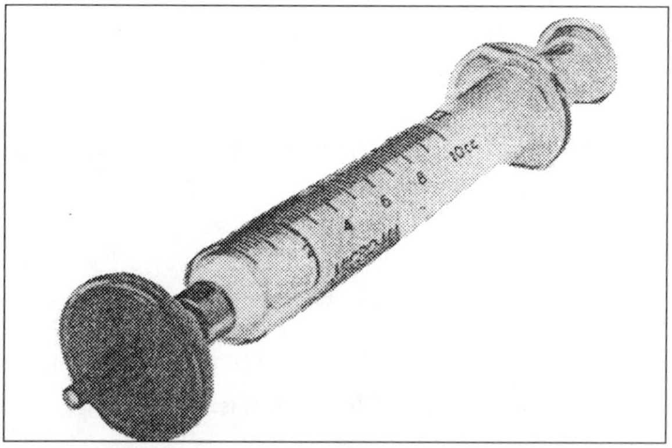

Fig. 5.2 Filtration of solvent through a nylon membrane of 0.13 mm diameter using a Syringe

most common deuterated solvents are deuterated chloroform ($CDCl_3$), deuterated acetone (CD_3COCD_3), deuterated benzene (C_6D_6).

5.5.1 Deutero-Chloroform ($CDCl_3$)

Method 1

Step (i) Preparation of anhydrous chloral from chloral hydrate

The chloral hydrate (CCl_3 CHO. H_2O) is treated with conc. sulphuric acid (20 ml for 40 g chloral hydrate). The mixture is allowed to stand for 1 hr at room temperature and then gently warmed for about 2 hr. Finally the mixture is distilled under anhydrous conditions, chloral is collected at 90-94°. The product is refractionated to get pure chloral (B.P. 90-94°/atm pressure).

Step (ii) Preparation of deutero chloroform from chloral

The reaction is carried out under anhydrous conditions under an atmosphere of nitrogen. To D_2O (22.5 ml) taken in a R.B flask is added freshly cut sodium pieces (5.25g) slowly. The flask is cooled in ice-cold water during the addition of sodium. Adequate precaution is taken to ensure that moisture does not get into the flask. The mixture is stirred and chloral (22.5 ml) is added drop-wise (during 30 minutes). Stirring is continued for 30 minutes more. During the above operation the temperature of the reaction mixture is kept below 0°. The mixture is stirred for 30 minutes at room temperature and then distilled using a small fractionating column. The fraction boiling between 52-54° is collected. Yield 36 ml. It is refractionated and in fraction 56-57° is collected.

Method 2

Deutrochloroform is conveniently prepared by the action of D_2O on hexachloroacetone in presence of pyridine.

$$CCl_3-\overset{\overset{\displaystyle O}{\|}}{C}-CCl_3 + D_2O \xrightarrow{\text{pyridine}} 2CDCl_3 + CO_2$$

Hexachloroacetone is commercially available. Dry hexachloroacetone and pyridine are used. The deutrochloroform is isolated as in the above method by fractional distillation.

5.5.2 Deutroacetone (CD_3COCD_3)

It is obtained by direct exchange process. Anhydrous acetone and D_2O are taken in the molar ratio (1:3). The solution is stirred after the addition of anhydrous potassium carbonate (catalyst). Stirring is continued for 4 hr.

No. of equilibrations	*Isotopic purity of acetone (%) (D)*
After 1st equilibration	74.9
After 2nd equilibration	92.25
After 3rd equilibration	97.45
After 4th equilibration	99.15

After each equilibration the mixture of acetone and deutroacetone is separated by filtration (under anhydrous conditions) and equilibrated again. Finally deutroacetone is obtained by fractional distillation.

5.5.3 Deutrobenzene (C_6D_6)

Step (i) Preparation of DCl gas by reacting thionyl chloride with heavy water

$$SOCl_2 + D_2O \rightarrow SO_2 + 2\ DCl$$

Step (ii) Reaction of DCl with benzene in presence of anhydrous aluminium chloride

Pure dry benzene and anhydrous aluminium chloride (powdered) is used. The reaction is carried out under anhydrous conditions.

$$C_6H_6 + DCl \xrightarrow{AlCl_3} C_6D_6 + HCl$$

The deutro benzene is isolated by fractional distillation (isotopic purity > 98%)

5.5.4 Deuterated Dimethyl Sulfoxide (DMSO - d_6)

Method (i) It is prepared by direct exchange method.

$$(CH_3)_2\ S{=}O + 3\ D_2O \xrightarrow{OH^-} (CD_3)_2\ S = O + 3H_2O$$

The catalyst used is K_2CO_3. Difficulty is encountered in this method to remove the last traces of water as DMSO gets solvated by water molecules. Calcium hydride is the best to remove water. But CaH_2 causes depletion as it undergoes deuteration.

$$CaH_2 + 2D_2O \rightarrow CaOHOD + HD + D_2$$

Hence it is necessary to use deuterated water.

Method (ii)

Step 1

Dimethyl sulphoxide is refluxed for 65 hr with excess methyl iodide (Normally 200 g of methyl iodide is used for 1000 g DMSO).

$$(CH_3)_2S{=}O + \underset{\text{(excess)}}{CH_3I} \xrightarrow[65\text{hr}]{\text{Reflux}} \underset{\text{(m.p. 200°C)}}{(CH_3)_3S^+OI^-}$$

The product $(CH_3)_3S^+OI^-$ is isolated by filtration.

Step-2

The $(CH_3)_3S^+OI^-$ is heated with D_2O at 80-100° in presence of anhydrous K_2CO_3 and then for 20 min at 200°.

$$(CH_3)_3S^+OI^- + 4\tfrac{1}{2}\ D_2O \xrightarrow[\text{anhyd. } K_2CO_3]{80-100°} CD_3\ SOI + 4\tfrac{1}{2}\ H_2O$$

$$CD_3\ SO1 \xrightarrow[20\text{ min}]{200°} (CD_3)_2\ S = O + CD_3\ I$$

DMSO-d_6 contaminated with iodine is purified by refluxing the compound with magnesium and distilling under reduced pressure.

This method has the special advantage that DMSO-d_6 is obtained pure and in good yield. Also deutrated chloroform ($CDCl_3$) is obtained as a byproduct.

5.6 PURIFICATION OF REAGENTS

5.6.1 Dimethyl Sulphate B.P. 188.5°

It is purified by washing with an equal volume of ice-cold water, followed by one third of the volume of ice-cold saturated sodium bicarbonate solution. It is finally dried over anhydrous calcium oxide. Final distillation gives pure dimethyl sulphate, B.P. 188.5°. It is advisable to distil under reduced pressure, B.P. 72-75°/13 mm. It should be stored in coloured stoppered bottles.

Caution: Dimethyl sulphate (both in vapour and liquid form) is highly poisonous. The liquid is readily absorbed through the skin with toxic results. If the liquid is accidentally splashed on the skin, wash immediately with excess of ammonia solution to hydrolyse the compound before it can be absorbed through the skin; finally rub gently with cotton soaked in ammonia solution.

5.6.2 Polyphosphoric Acid

It is prepared by dissolving 1.8 parts by weight of phosphorus pentoxide in 1 part by weight of orthophosphoric acid (H_3PO_4); the mixture is heated in a boiling water bath for 20 min and immediately used for the reaction.

5.6.3 Sodium Pulverised

Clean sodium metal (required amount) is taken in a conical flask (150 ml capacity) and covered with dry xylene. The flask is heated on a sand bath till the vapours of xylene come near to the neck of the flask. The flask is removed from the sand bath, stoppered with a rubber stopper, covered with a cloth and shaken vigorously; the pressure is released after shaking. The xylene is decanted and pulverised sodium is washed twice with dry ether. It is directly used for the reaction.

5.6.4 Sodium Acetate Fused

It is heated in a porcelain dish over a small flame using a wire gauze and a glass rod for stirring. It liquefies, steam is evolved and the mass solidifies as soon as most of the water of crystallisation has been removed. The heating is continued till the solid just melts. Excessive heating is avoided. The fused salt is allowed to solidify in the dish and is removed while still warm. It is powdered and kept in tightly stoppered bottles.

5.6.5 Sodium Amalgam

Clean sodium (11.8 g) is placed in a three necked R.B. flask (250 ml capacity) fitted with a dropping funnel, an inlet for nitrogen gas and an outlet tube. The air is displaced by nitrogen. Mercury (5 ml) is added and the reaction flask is warmed gently with a free flame or until the reaction commences. The flame is removed and the reaction is maintained by slow addition of mercury. When approximately half the mercury has been added the amalgam tends to solidify. It is kept molten by heating with occasional shaking. After addition of all mercury (total 375 g), the hot amalgam is poured in a mortar and powdered. It is kept in tightly stoppered bottle. The last step of powdering is performed in a dry cupboard under nitrogen atmosphere. The sodium amalgam thus obtained contains 1.2% sodium.

5.6.6 Thionylchloride B.P. 77°

Commercial thionylchloride is distilled over quinoline to remove acidic impurities (10 ml quinoline for 100 g thionylchloride).

5.6.7 Zinc Amalgam

A mixture of zinc wool (200 g), mercuric chloride (15 g), concentrated hydrochloric acid (10 ml) and water (250 ml) is shaken for 5-10 min. The aqueous solution is decanted and the amalgamated zinc is covered with water (150 ml) and concentrated hydrochloric acid (200 ml). The material (0.3-0.4 mole) to be reduced by Clemmensen reduction is then added immediately and the reaction carried out.

5.6.8 Zinc Chloride Fused

Zinc chloride is heated in an porcelain dish over a free flame using a wire gauze. It melts after some time. The molten zinc chloride is stirred gently with a glass rod. When it starts fuming, it is poured in a dry mortar, powdered and kept in a tightly stoppered bottle. The last step of powdering should be done in a shortest possible time, preferably in a cupboard free of moisture.

Bibliography

1. J.A. Riddick and W.B. Burger, *Techniques of Chemistry*, 3rd Edn., Ed. A. Weissberger, New York: Wiley Interscience, Vol. II, 1970.
2. L.F. Fieser and M. Fieser, *Reagents for Organic Synthesis*, New York, Wiley Interscience, Vols. 1-5.
3. D.D. Perrin, W.L.F. Armarego and D.K. Perrin, *Purification of Laboratory Chemicals*, Oxford, Pergamon Press, 1966.